```
┌─────┐
│ PBM │
└─────┘
  ↓      ↓       ↓
┌───────┐┌──────┐┌─────┐
│PHYSICS││BRAIN ││ Map │
└───────┘└──────┘└─────┘
```

Quick Revision
Chapterwise Formula's
Physics Brain Map
Vol. 1

Class: 11th

Formulae extracted from each lines of NCERT Book.

<u>Written By:</u>

Azhar Mahmood Mir

PREFACE

The motto underlying the book is physics is enjoyable. I have tried to use symbols, names, etc. which are popular nowaday.

I have tried my best to keep erros outs of this book. I shall be gratefull to the readers who point out any errors and/or make other constructive suggestions.

Readers are requested to send their queries, suggestions. This will help me to improve and serve you better. Kindly e-mail your views to mirazhar103@gmail.com

"A great deal of my work is just playing with equation and seeing what they give."

To The Students:

The book contain 29 chapters divided in two parts (such as Part-1 contain 14 chapters, in which all important formulae strictly based on NCERT and some extra materials). (And Part-II contain 15 chapters of class 12th. In which all important formulae are written).

As per current trend, most of the questions asked directly from the formula in NEET as well as JEE.

Now you can easily remember all the formulae of the Physics. In this book only formulae written, which help you to revise whole Physics book within almost 2 weeks for the competitive exam such as NEET and JEE.

"To get to know
to discover, to
publish — this is
the destiny of a
scientist."

Contents:

EDITION – 2022

1. Units and Measurement
2. Motion in a Straight Line
3. Motion in a Plane
4. Laws of Motion
5. Work, Enery and Power
6. System of Particles and Rotational Motion
7. Gravitation
8. Mechanical Properties of Solids
9. Mechanical Properties of Fluids
10. Thermal Properties of Matter
11. Thermodynamics
12. Kinetic Theory
13. Oscillations
14. Waves.

UNITS AND MEASUREMENT

- Length → metre
- Mass → Kilogram
- Time → Second
- Temperature → Kelvin
- Electric Current → Ampere
- Luminous intensity → Candela
- Amount of Substance → mole

⇒ Convert a Physical Quantity from one system of units to another.

Physical Quantity + Units

$$n_2 = n_1 \left(\frac{u_1}{u_2}\right) = n_1 \left(\frac{M_1}{M_2}\right)^a \left(\frac{L_1}{L_2}\right)^b \left(\frac{T_1}{T_2}\right)^c$$

- n_2 → numerical value of I system
- n_1 → numerical value of I system
- M_1 → unit of mass in I system
- M_2 → unit of mass in II system
- L_1 → unit of length in I system
- L_2 → unit of length in II system
- T_1 → unit of time in I system
- T_2 → unit of time in II system

Rules For Significant Figures:

1. All non-zero digits are significant figures.

   e.g., 49 has <u>2SF</u>

2. All zeros occurring b/w non-zero digits are significant figures.

   e.g., 403 has <u>3SF</u>

3. All zeros to the right of the last non-zero digit are not significant figures. e.g. 20% has <u>3SF</u>

4. All zeros to the right of a decimal point and to the left of a non-zero digit are not significant figures.

   e.g., 0.09 has <u>1SF</u>.

5. All zeros to the right of a decimal point and to the right of a non-zero digit are significant figures.

   e.g., 0.30 has <u>2SF</u>

ERRORS IN MEASUREMENTS

→ Rounding off the measurements :-

1. If the digit to be dropped in a number is less than 5, then the preceeding digit remains unchanged.

   e.g. 8.64 is rounded off to 0.6

   [ERRORS IN MEASUREMENTS]

2. If the digit to be dropped in a number is greater than 5, then the preceeding digit is raised by 1.

   e.g. 7.66 is rounded off to 7.7

3. If the digit to be dropped in a number is 5 followed by zeros, then the preceeding digit is

remains unchanged if it is even.

e.g.: 0.65 is rounded off to 0.6

" : 0.650 is rounded off to 0.6.

4. If the digit to be dropped in a number is 5 or 5 followed by zeros, then the preceding digit is raised by 1 is it is odd.

e.g.: 0.75 is rounded 27 to 0.8

ERRORS IN MEASUREMENT

ERRORS IN MEASUREMENTS

→ **1. Absolute error (Δa_i)**

$$= \text{True Value} - \text{Measured Value}$$

$$\Delta a_i = \overline{a} - a_i$$

$\overline{a} \rightarrow$ Mean value of the measured quantity.

$a_i \rightarrow$ Value of the quantity measured in i^{th} observation.

→ **Mean Absolute error ($\Delta \overline{a}$)**

$$\Delta \overline{a} = \frac{|\Delta a_1| + |\Delta a_2| + \dots}{n}$$

$\Delta a_1 = \overline{a} - a_1$

$\Delta a_2 = \overline{a} - a_2$

→ **Relative Error:**

$$\delta a = \frac{\Delta \overline{a}}{\overline{a}}$$

→ **Percentage Error:** $\delta a \times 100\% = \frac{\Delta \overline{a}}{\overline{a}} \times 100\%$

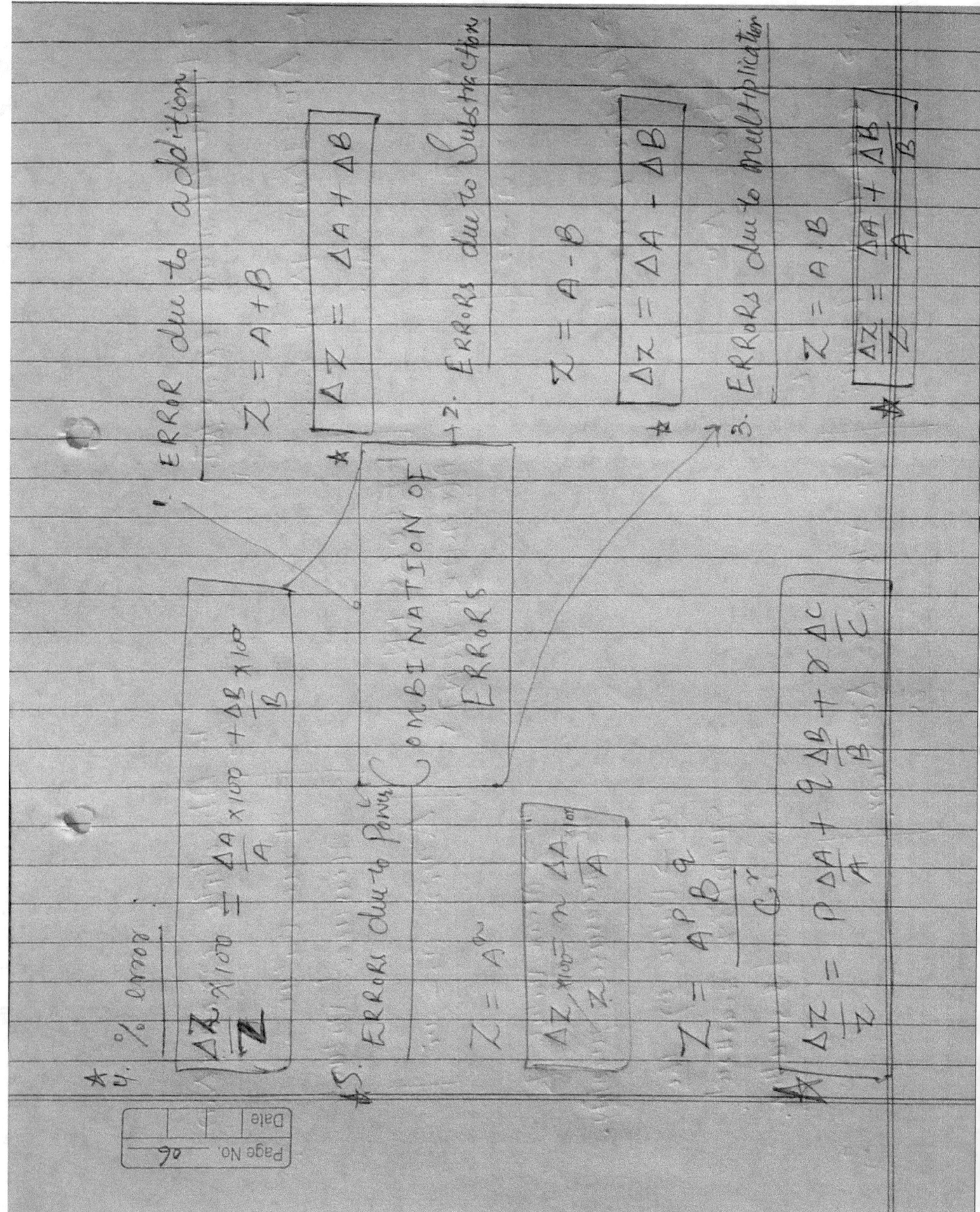

COMBINATION OF ERRORS

1. Error due to addition

$Z = A + B$

$$\boxed{\Delta Z = \Delta A + \Delta B}$$

2. Errors due to Substraction

$Z = A - B$

$$\boxed{\Delta Z = \Delta A - \Delta B}$$

3. Errors due to Multiplication

$Z = A \cdot B$

$$\boxed{\dfrac{\Delta Z}{Z} = \dfrac{\Delta A}{A} + \dfrac{\Delta B}{B}}$$

★ ⁂. % error:

$$\boxed{\dfrac{\Delta Z}{Z} \times 100 = \dfrac{\Delta A}{A} \times 100 + \dfrac{\Delta B}{B} \times 100}$$

5. Errors due to Power

$Z = A^n$

$$\boxed{\dfrac{\Delta Z}{Z} \times 100 = n \dfrac{\Delta A}{A} \times 100}$$

$Z = \dfrac{A^p B^q}{C^r}$

★ $$\boxed{\dfrac{\Delta Z}{Z} = p\dfrac{\Delta A}{A} + q\dfrac{\Delta B}{B} + r\dfrac{\Delta C}{C}}$$

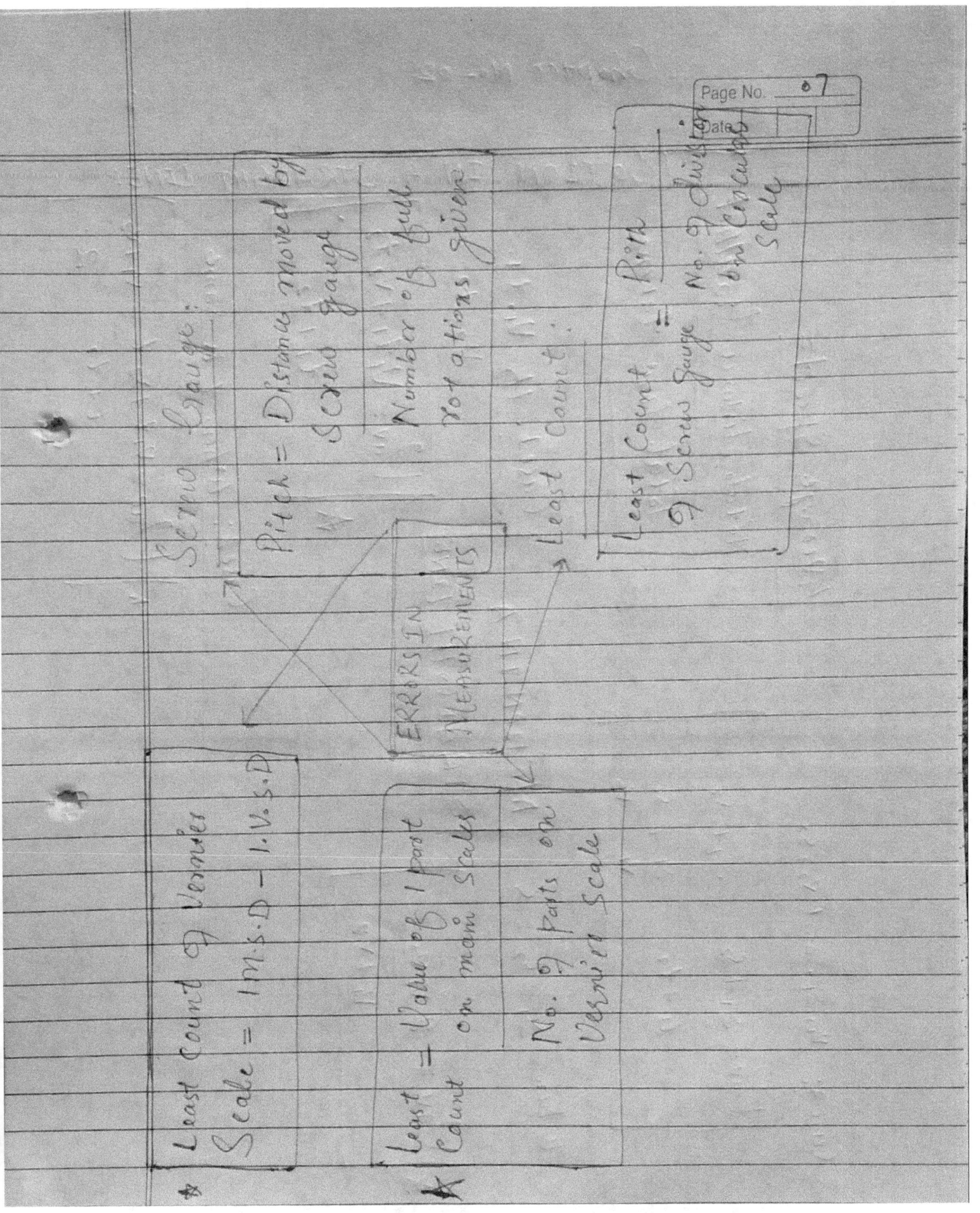

※ Least Count of Vernier
$$\text{Scale} = 1 M.S.D - 1.V.S.D$$

$$\text{Least Count} = \frac{\text{Value of 1 part on main scale}}{\text{No. of parts on Vernier Scale}}$$

● Screw Gauge:

$$\text{Pitch} = \frac{\text{Distance moved by Screw gauge}}{\text{Number of full rotations given}}$$

→ Least Count:

$$\text{Least Count of Screw Gauge} = \frac{\text{Pitch}}{\text{No. of divisions on circular scale}}$$

ERRORS IN MEASUREMENTS

MOTION IN A STRAIGHT LINE:

Average Speed = $\dfrac{\text{Total distance travelled}}{\text{Total time taken}}$

MOTION IN A STRAIGHT LINE

Case 1: if v_1, t_1, v_2, t_2

$$\boxed{V_{av} = \dfrac{v_1 t_1 + v_2 t_2}{t_1 + t_2}}$$

★

Case 2: when t is same

$$\boxed{V_{av} = \dfrac{v_1 t + v_2 t}{t + t} = \dfrac{v_1 + v_2}{2}}$$

★

Case 3: $v_1, S_1,$ and v_2, S_2

$$\boxed{V_{av} = \dfrac{S_1 + S_2}{t_1 + t_2} = \dfrac{S_1 + S_2}{\dfrac{S_1}{v_1} + \dfrac{S_2}{v_2}}}$$

when $S_1 = S_2 = S$

$$\boxed{V_{av} = \dfrac{2 v_1 v_2}{v_1 + v_2}}$$

★

MOTION IN A STRAIGHT LINE

Equations of Motion

$$v = ut + \frac{1}{2}at^2$$

$$v^2 = u^2 + 2as$$

$$S = ut + \frac{1}{2}at^2$$

$$D_n = u + \frac{a}{2}(2n-1)$$

★ $\dfrac{D_n}{S_n} = \dfrac{2n-1}{n^2}$

★ $t = \sqrt{\dfrac{2h}{g}}$

★ $v = \sqrt{2gh}$

★ $t_{up} = u/g$

★ $t_{down} = u/g$

★ $h_{max} = u^2/2g$

★ $\dfrac{t_{ac}}{t_{descent}} = \sqrt{\dfrac{g-a_0}{g+a_0}}$

→ Relative velocity

★ $\vec{V}_{AB} = \vec{V}_A - \vec{V}_B$

Special Cases :

→ When the object A and B move in the same direction

★ $V_{AB} = V_A - V_B$

→ When the object A and B move in opposite direction.

★ $V_{AB} = V_A + V_B$

MOTION IN A STRAIGHT LINE

Chapter No - 03

MOTION IN A PLANE:

||gm law:
$$\vec{R} = \sqrt{A^2 + B^2 + 2AB\cos\theta}$$

△ law:
$$\vec{R} = \sqrt{A^2 + B^2 + 2AB\cos\theta}$$

Polygon law:
$$\vec{R} = \vec{A} + \vec{B} + \vec{C} + \vec{D}$$

MOTION IN A PLANE

⇒ **Lami's Theorem:**

$$\frac{|\vec{A}|}{\sin\alpha} = \frac{|\vec{B}|}{\cos\beta} = \frac{|\vec{C}|}{\sin\gamma}$$

Scalar/Dot Products

$$\vec{A}\cdot\vec{B} = |\vec{A}||\vec{B}|\cos\theta$$

$$\hat{i}\cdot\hat{i} = \hat{j}\cdot\hat{j} = \hat{k}\cdot\hat{k} = 1$$

$$\hat{i}\cdot\hat{j} = \hat{j}\cdot\hat{k} = \hat{k}\cdot\hat{i} = 0$$

$\vec{A} \perp \vec{B}, \quad \theta = 90°$
$\vec{A}\cdot\vec{B}, \quad \theta = 0°$

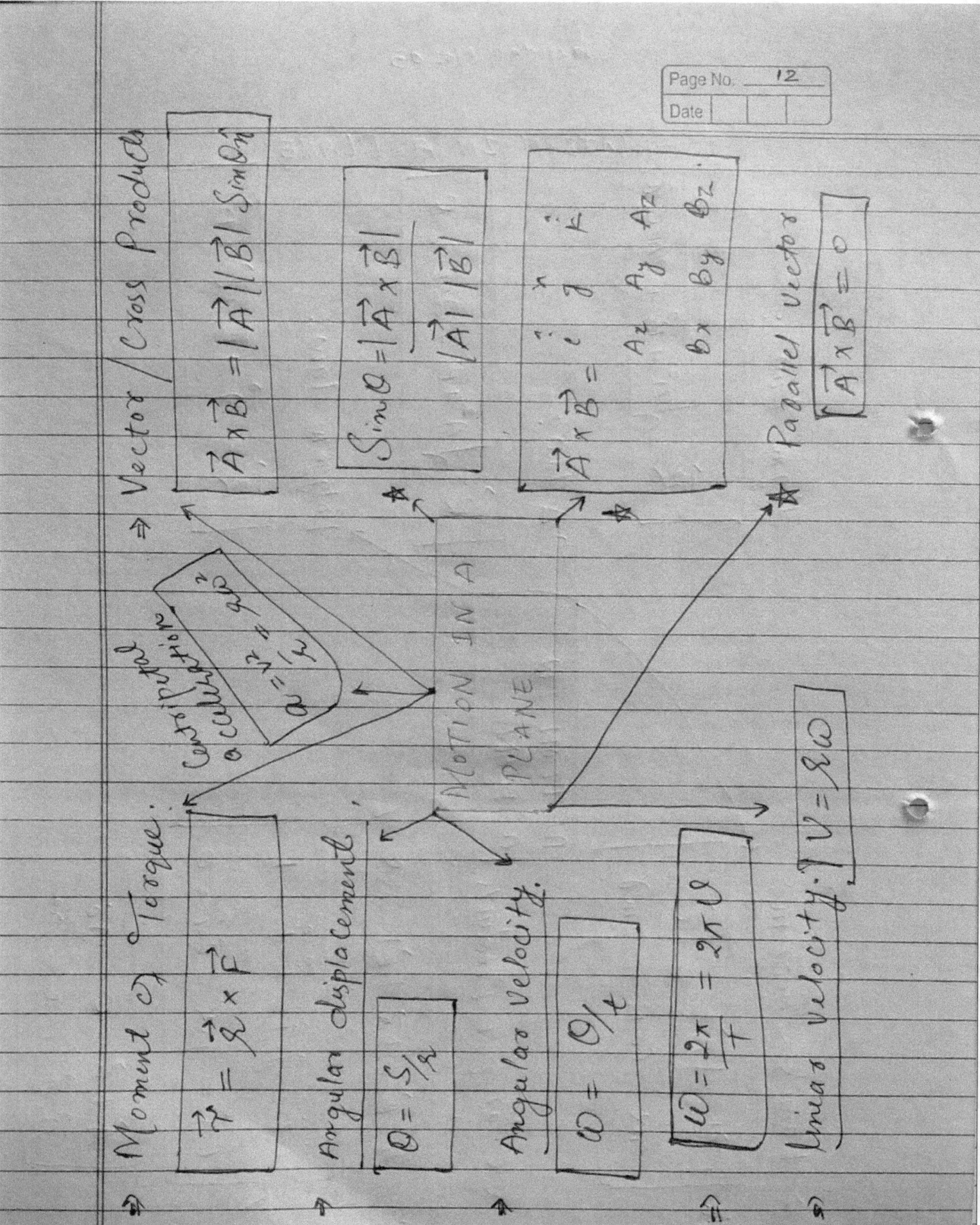

MOTION IN A PLANE

→ Equation of Trajectory:

$$y = x\tan\theta - \frac{g \cdot x^2}{2u^2\cos^2\theta}$$

- Maximum height:

$$H = \frac{u^2\sin^2\theta}{2g}$$

- Horizontal Range:

$$R = \frac{u^2\sin 2\theta}{g}$$

→ Max. horizontal Range:

$$R_{max} = \frac{u^2}{g}$$

$$V = \sqrt{V_x^2 + V_y^2}$$

$$\tan\beta = \frac{V_y}{V_x}$$

MOTION IN A PLANE

Time of flight
$$T = \dfrac{2u\sin\theta}{g}$$

Banking of Roads
$$\tan\theta = \dfrac{v^2}{rg}$$

$$U_1 \sin\theta_1 = U_2 \sin\theta_2$$

Change in Momentum
$$\vec{\Delta P} = \vec{P_f} - \vec{P_i}$$

$$V_{max} = \sqrt{rg\tan\theta}$$

Chapter No- 04

LAW OF MOTION

- **Momentum:**

$$\vec{P} = m\vec{v}$$

→ Unit → kg-m/s or slug-m/s

→ D.F → $M^1L^1T^{-1}$

LAW OF MOTION

- **Newton's Second law of Motion:**

☆ $$\vec{F} = \frac{d\vec{P}}{dt} = \frac{d m\vec{v}}{dt}$$

☆ $$\vec{F} = m\frac{d\vec{v}}{dt} = m\left(\frac{v-u}{t}\right)$$

- **Impulse (J):** SI unit → N-s or kg-m/s

$$\vec{J} = F \times t$$

☆ $$\vec{J} = \int_{t_1}^{t_2} \vec{F} \cdot dt$$

D.F → $M^1L^1T^{-2}$

$$\vec{J} = F \times \Delta t = m(v-u)$$

- Average Retarding force.

$$F = \frac{Impulse}{Time}$$

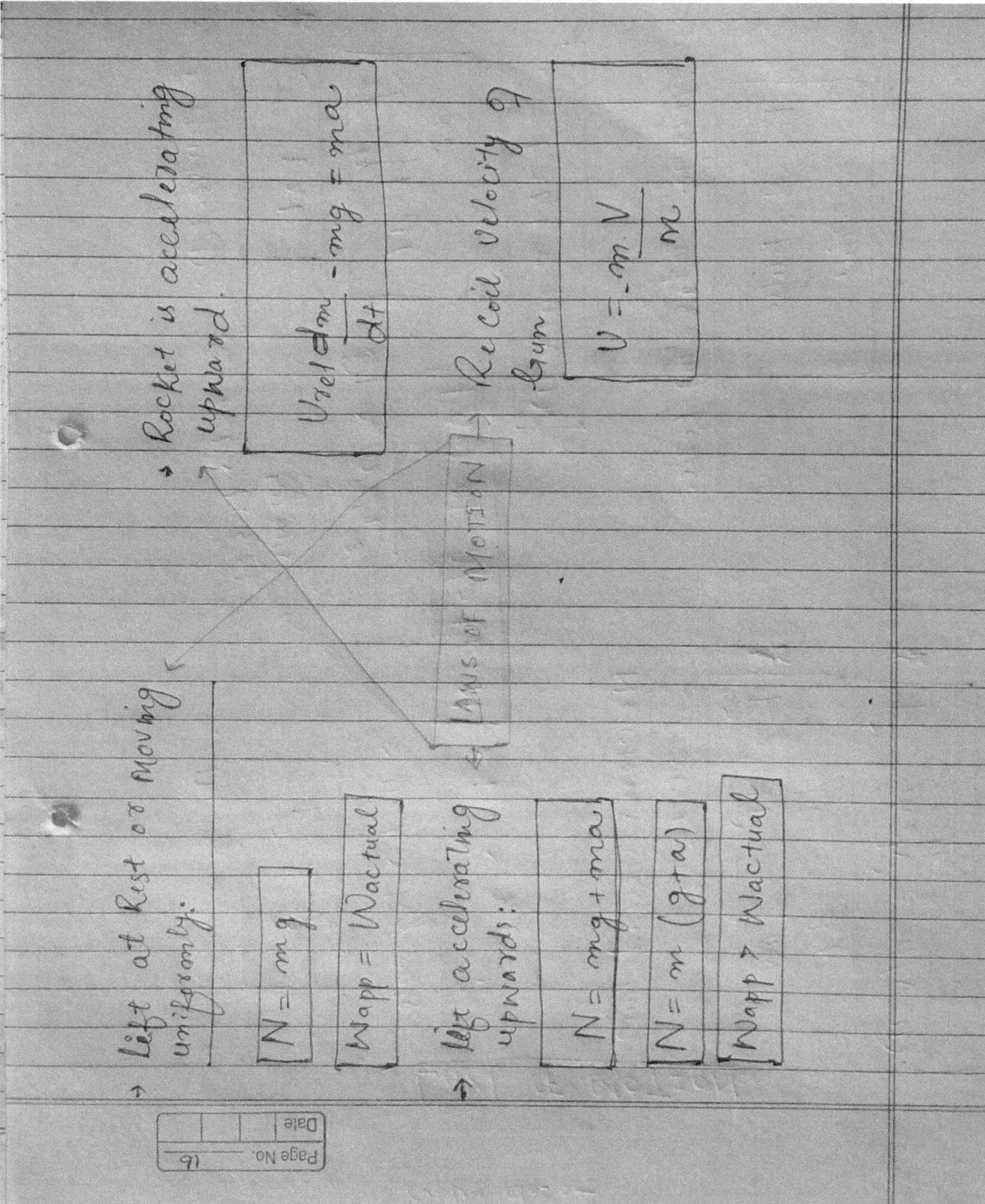

LAWS OF MOTION

→ Rocket is accelerating upward.

$$V_{rel}\frac{dm}{dt} - mg = ma$$

→ Recoil velocity of Gun

$$V = -m\frac{V}{m}$$

→ Lift at Rest or Moving uniformly:

$$N = mg$$

$$W_{app} = W_{actual}$$

→ Lift accelerating upwards:

$$N = mg + ma$$
$$N = m(g+a)$$

$$W_{app} > W_{actual}$$

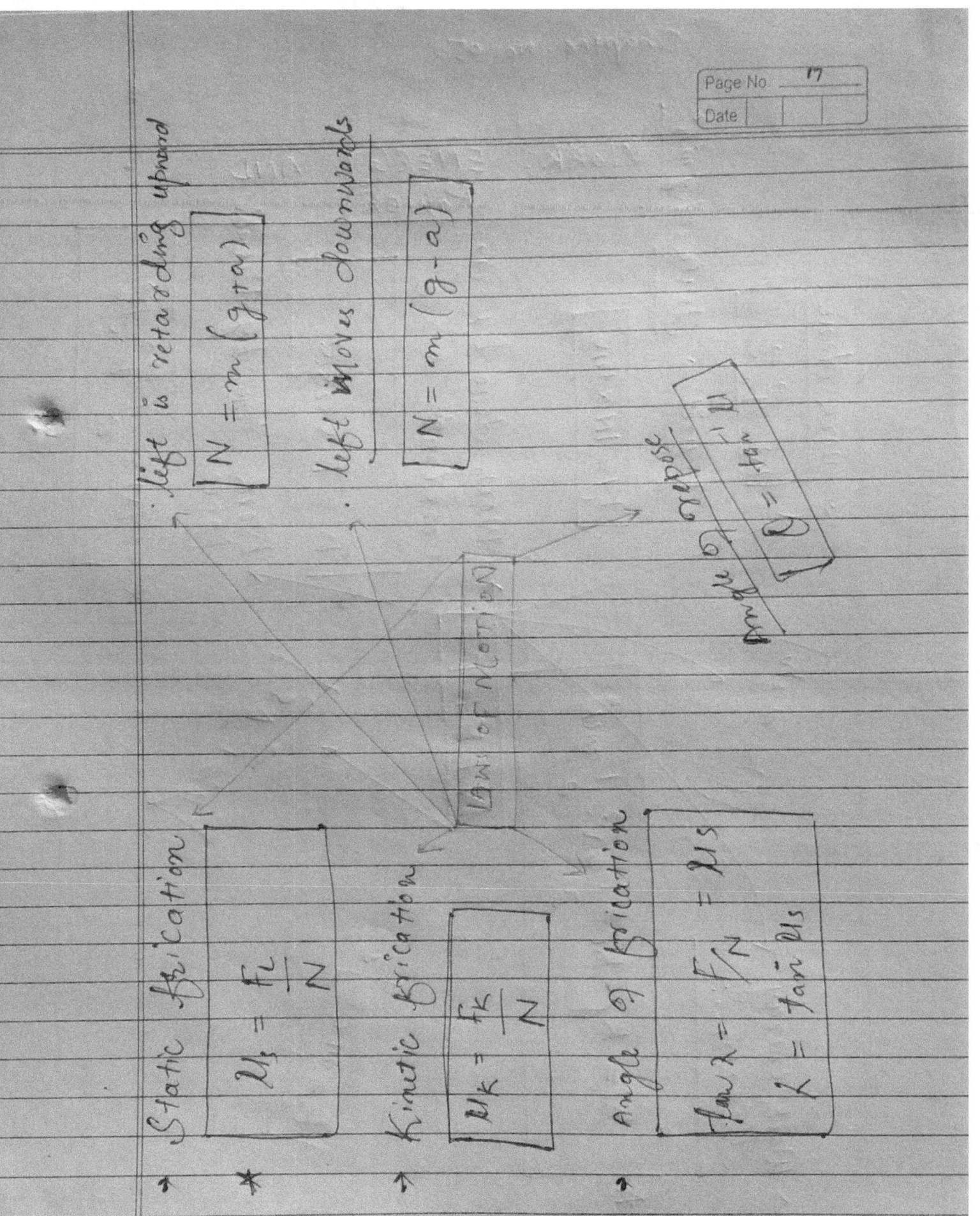

Chapter No. 05

WORK, ENERGY AND POWER

→ Work done:
$$\vec{W} = \vec{F} \cdot \vec{S}$$

SI unit → Joule / $kg \, m^2/s^2$

$$1J = 1N \times 1m$$

→ Work done by a Constant force
$$W = FS \cos\theta$$

→ Work done by Variable force
$$W = \int_{i}^{f} \vec{F} \cdot d\vec{r}$$

$$W = mgh$$

→ Work energy Theorem:
$$W = K_f - K_i = \frac{1}{2}mv^2 - \frac{1}{2}mu^2$$

$$W + E + P$$

$$W = \int \vec{F} \cdot d\vec{r}$$

$$E = mc^2$$

⇒ Kinetic energy of a body A is by M%, Then %↑ in Momentum

⇒ Kinetic Energy
$$K.E = \frac{1}{2}mv^2$$

$$F = -\frac{du}{dx}$$
$$\Rightarrow du = -F\,dx$$

→ Work done by Spring force
$$W = \frac{1}{2}k(x_2^2 - x_1^2)$$

%↑ in Momentum
$$= \left(\left(\sqrt{\frac{100+n}{100}}\right) - 1\right) \times 100 \quad \boxed{NHS + 9}$$

%↑ in Kinetic Energy
$$= \left(\left(\frac{100+n}{100}\right)^2 - 1\right) \times 100$$

- Hook's law:

$$F = -kx$$

- Force Constant

$$k = F/x$$

- Potential energy of Spring.

$$U = \frac{1}{2}kx^2$$

$$W + E + P$$

- Ratio of Potential energy Stored:

$$\frac{U_1}{U_2} = \frac{k_2}{k_1}$$

- Ratio of K.E.:

$$\frac{K.E_1}{K.E_2} = \frac{m_2}{m_1}$$

Power:
$$P = \dfrac{W}{t} = \dfrac{\Delta W}{\Delta t}$$

$$P = \vec{F} \cdot \vec{v}$$

$$P = v^2 \dfrac{dm}{dt}$$

$$E_T = \dfrac{P^2}{2m}$$

$$W + E = P$$

$$\dfrac{k_2}{k_1} = \dfrac{P_2^2}{P_1^2}$$

★ Efficiency:
$$\eta = \dfrac{\text{output Power}}{\text{input Power}}$$

★ Output power $= \dfrac{mgh}{t}$

• Ratio of Momentum:
$$\dfrac{P_1}{P_2} = \sqrt{\dfrac{m_1}{m_2}}$$

→ Coefficient of restitution for a collision is given by:

$$e = \frac{v_1 - v_2}{u_1 - u_2} = \frac{|v_2 - v_1|}{|u_1 - u_2|}$$

★

$KE = E + P$

→ Ball rebounding from a floor,

$$e = v/u$$

- For elastic collision

$$e = 1$$

For inelastic collision

$$e < 1$$

Chapter No - 06

SYSTEM OF PARTICLES AND ROTATIONAL MOTION

→ **C.O.M of System of Point Masses:**

$$X_{cm} = \frac{m_1 x_1 + m_2 x_2 + \dots + m_n x_n}{m_1 + m_2 + \dots + m_n}$$

$$\vec{r}_{cm} = \frac{m_1 \vec{r}_1 + m_2 \vec{r}_2 + \dots + m_n \vec{r}_n}{m_1 + m_2 + \dots + m_n}$$

→ **Velocity:**

★ $\vec{V}_{cm} = \dfrac{m_1 \vec{v}_1 + m_2 \vec{v}_2 + \dots + m_n \vec{v}_n}{m_1 + m_2 + \dots + m_n}$

→ **Torque:** $\vec{\tau} = \vec{r} \times \vec{F}$

→ **Power of Torque:**

$$P = \tau \omega$$

→ **Work done by a Torque:**

$$W = \tau \theta$$

→ **Angular Momentum:**

$$L = r p \sin\theta \quad \text{or} \quad \vec{L} = \vec{r} \times \vec{p}$$

	Inertia	β
1. Ring	MR^2	β = 2
2. Hollow cylinder	MR^2	β = 2
1. Disc	$\frac{1}{2}MR^2$	β = 3/2
2. Solid cylinder	$\frac{1}{2}MR^2$	β = 3/2
1. Hollow Sphere	$\frac{2}{3}MR^2$	β = 5/3
2. Solid Sphere	$\frac{2}{5}MR^2$	β = 7/5

System of Particles And Rotational Motion

☆ 1. $(KE)_{rolling} = β \times \frac{1}{2}mV_{cm}^2$

Rolling Motion:

- $V_{cm} > r\omega \rightarrow$ Forward Slipping
- $V_{cm} < r\omega \rightarrow$ Backward Slipping
- $V_{cm} = r\omega \rightarrow$ pure Rolling (no slipping)

- Parallel Axis Theorem

$$I_z = I_x + I_y$$

- Perpendicular Axis Theorem

$$I = I_{cm} + md^2$$

System of Particles and Rotational Motion

$$V_{cm} = \sqrt{\frac{2gh}{\beta}}$$

$$t = \frac{1}{\sin\theta}\sqrt{\frac{2hx}{g}}$$

$$F_T = \frac{1}{\beta}$$

$$F_{rot} = 1 - \frac{1}{\beta}$$

$$a_{cm} = \frac{g\sin\theta}{\beta}$$

Linear acceleration

$$a = \frac{2}{3} g\sin\theta$$

S.P.M

Rotational KE:

$$= \frac{1}{2} I \omega^2$$

Radius of Gyration:

$$I = mK^2$$

$$K = \sqrt{\frac{I}{m}}$$

Chapter No - 07

GRAVITATION

⇒ Universal Law of Gravitation:

$$F = \frac{G m_1 m_2}{r^2}$$

★ $G = 6.67 \times 10^{-8}$ dyne cm^2/g^2

$G = 6.67 \times 10^{-11}$ Nm^2/kg^2

→ SI unit → $M^{-1} L^3 T^{-2}$

⇒ Gravitational field intensity:

$$I = F/m$$

→ Mass of Earth:

$$M = \frac{gR^2}{G}$$

⇒ Acceleration due to Gravity:

$$g = \frac{GM}{R^2} = \frac{F}{m}$$

GRAVITATION

Due to altitude: $h < R$

$$g_h = g\left(1 - \frac{2h}{R_e}\right)$$

i) $\dfrac{\Delta g}{g} = \dfrac{\Delta M}{M} - \dfrac{2\Delta R}{R}$

ii) $g \propto \dfrac{1}{R^2}$

$\dfrac{\Delta g}{g} = -\dfrac{2\Delta R}{R}$

Due to depth:

$$g_d = g\left(1 - \frac{d}{R_e}\right)$$

i) $\dfrac{\Delta g_d}{g_d} = \dfrac{d}{R_e}$

Latitude:

$$g_\lambda = g - R\omega^2 \cos^2\lambda$$

$$g_{equ} = g - R\omega^2$$

→ Gravitational Potential Energy:

$$U = -\frac{GMm}{R}$$

SI unit → Joule

Df → $M^1 L^2 T^{-2}$

∴ Orbital Velocity:

$$V_o = \sqrt{\frac{Gm}{R_e}} = \sqrt{gR_e}$$

→ Kepler's Laws

1. Sun at a Focii (Law of orbits)

2. Law of Areas:

   Law of Conservation of GRAVITATION

   of Angular Momentum.

   * Areal Velocity

   $$\frac{dA}{dt} = \frac{L}{2m}$$

3. Third Law (Law of Periods)

   $$T^2 \propto R^3$$

   $$T = 2\pi \sqrt{\frac{R^3}{M}}$$

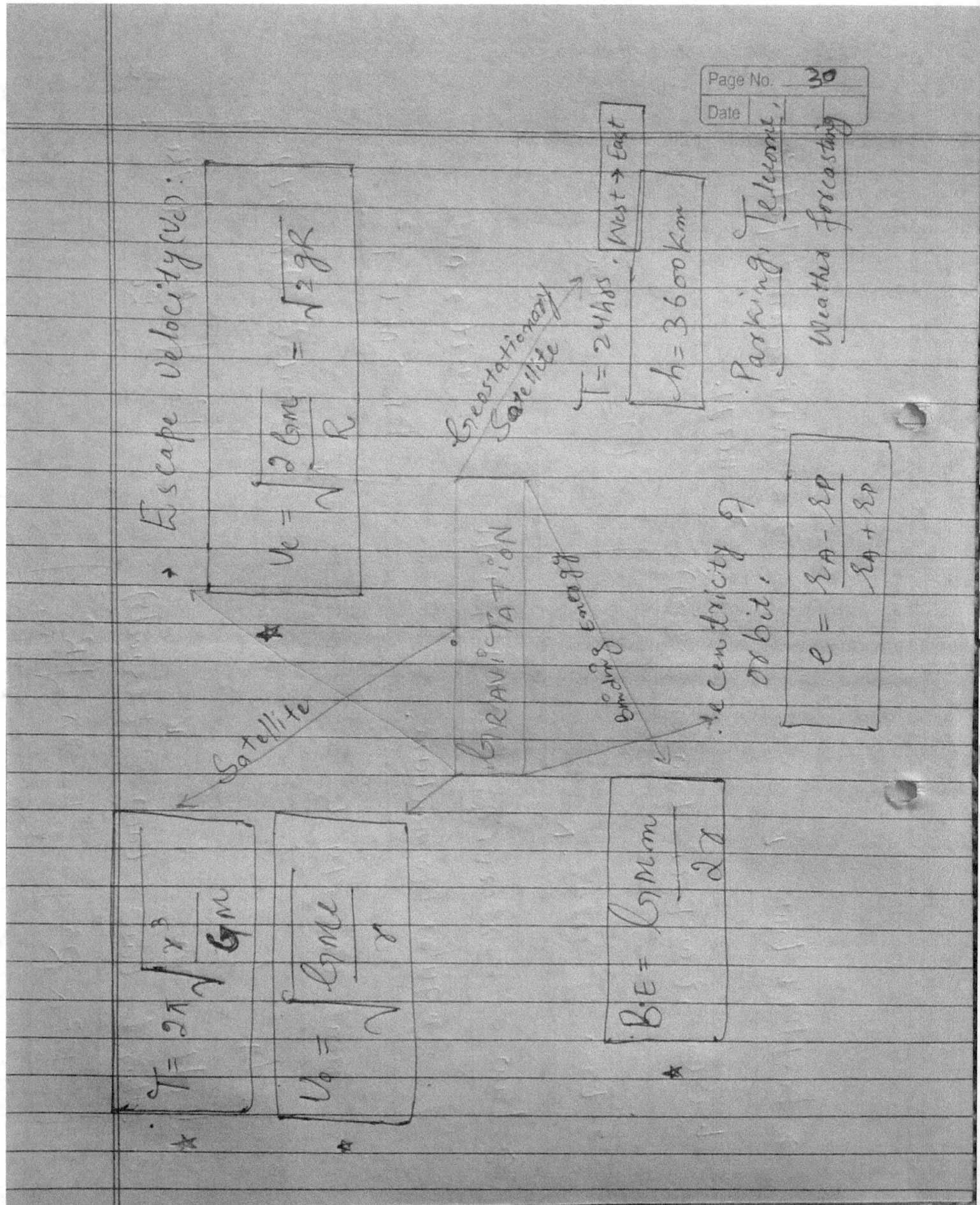

GRAVITATION

- **Escape Velocity (v_e):**
$$v_e = \sqrt{\frac{2Gm}{R}} = \sqrt{2gR}$$

- Satellite:
  $$T = 2\pi\sqrt{\frac{r^3}{Gm}}$$
  $$v_o = \sqrt{\frac{Gm_e}{r}}$$

- Binding Energy:
$$B.E = \frac{GMm}{2r}$$

- Geostationary Satellite: West → East
  $T = 24\,hrs$
  $h = 36000\,km$
  Parking, Telecom, Weather forecasting

- Eccentricity of orbit:
$$e = \frac{r_A - r_P}{r_A + r_P}$$

Chapter No - 08

MECHANICAL PROPERTIES OF SOLIDS

→ **Stress:**

$$\text{Stress} = \frac{\text{Force}}{\text{Area}} = \frac{F}{A}$$

★ D.M → $[M^1L^{-1}T^{-2}]$

SI unit → $[N\,m^{-2}]$

→ Longitudinal Strain

$$L.S = \frac{\text{Change in length}}{\text{original length}}$$

M.P.S

→ **Breaking Stress:**

$$B.S = F/A$$

⇒ **Hooke's law:**

Stress ∝ Strain

$$\frac{\text{Stress}}{\text{Strain}} = \text{Constant}$$

★ Total Potential energy stored in a stretched wire:

$$U = \frac{1}{2} F \cdot \Delta L$$

→ Young Modulus of elasticity

★ $Y = \dfrac{\text{Longitudinal Stress}}{\text{Longitudinal Strain}}$

$Y = F/A \Big/ \dfrac{\Delta L}{L} = \dfrac{FL}{\ell A}$

S.I unit → $[N/cm^2]$

D.M → $[M^1 L^{-1} T^{-2}]$

· Poisson's ratio:

$\sigma = \dfrac{\text{lateral strain}}{\text{longitudinal strain}}$

$\sigma = \dfrac{\Delta D/D}{\Delta \ell/\ell}$

★ % ↑ in length:

$\dfrac{\Delta \ell}{\ell} \times 100 = \dfrac{F}{AY} \times 100$

M.P.S

⇒ Compressibility (C):

$$C = \frac{1}{B} = -\frac{\Delta V}{PV}$$

⇒ Bulk's Modulus of Elasticity K or B:

$$B = \frac{F/A}{-\Delta V/V} = \frac{-\Delta PV}{\Delta V}$$

SI unit → N/m^2

M.P.S

⇒ Modulus of Rigidity η:

$$\eta = \frac{F/A}{\theta} = \frac{PL}{\Delta L}$$

Chapter No - 09

MECHANICAL PROPERTIES OF FLUIDS:

Pressure:

$$P = \frac{F}{A}$$

SI unit → Pascal

DF → $ML^{-1}T^{-2}$

→ Gauge Pressure.

$$P = P_a + P_g$$

P_a → absolute Pressure
P_g → gauge Pressure

Venturimeter:

$$Q = a_1 a_2 \sqrt{\frac{2gh}{a_1^2 - a_2^2}}$$

$$V = \sqrt{2gh}$$

M.P.F

Equation of Continuity:

$$AV = Constant$$

or $\quad A_1 V_1 = A_2 V_2$

Conservation of Mass

Bernoulli's Theorem:

• First Form:

$$\frac{P}{\rho} + gh + \frac{V^2}{2} = Constant$$

• 2nd form:

$$\frac{P}{\rho g} + h + \frac{V^2}{2g} = Constant$$

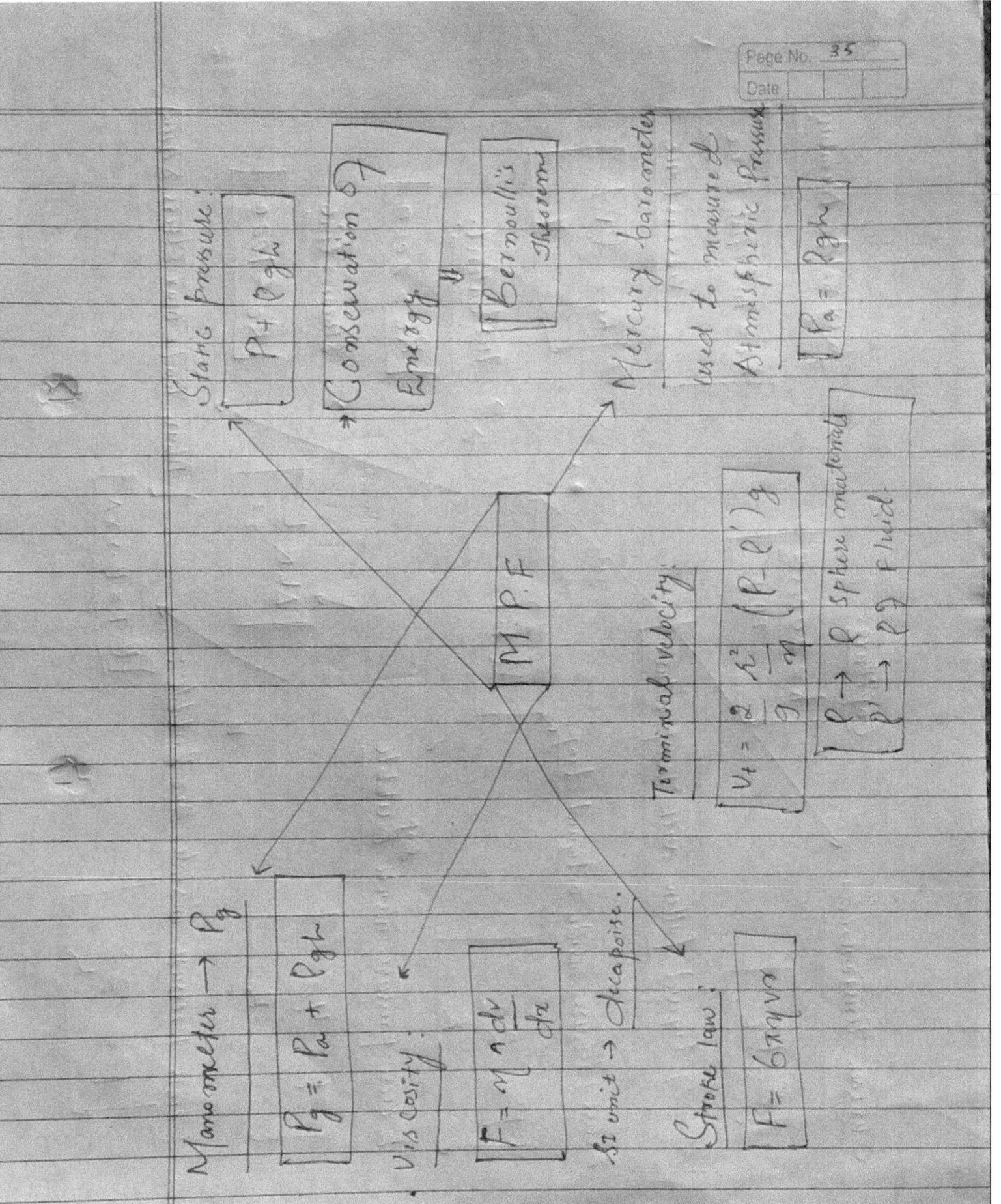

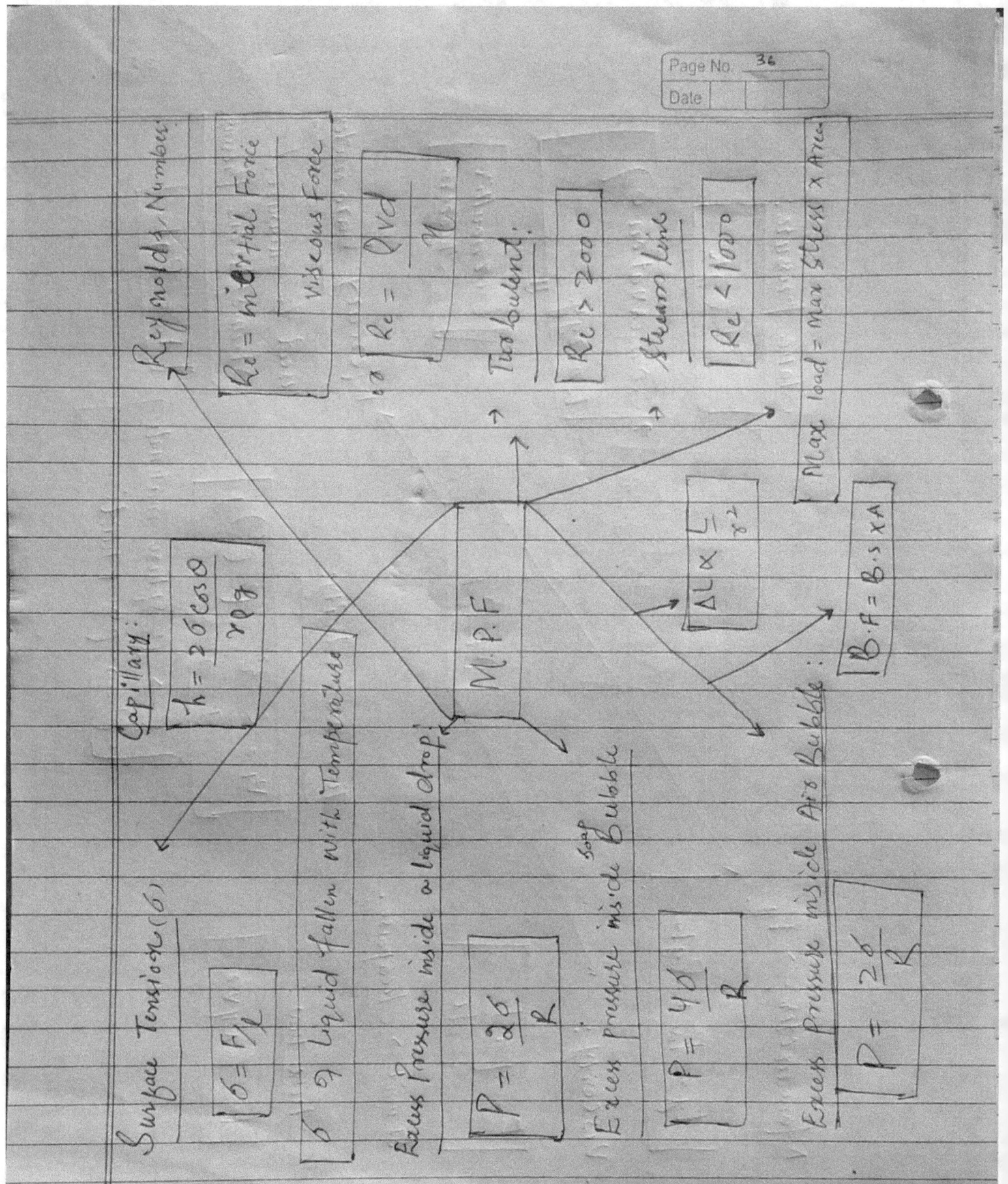

Reynolds Numbers

$$Re = \frac{\text{Inertial Force}}{\text{Viscous Force}}$$

$$Re = \frac{\rho v d}{\mu}$$

Turbulent: $Re > 2000$

Stream Line: $Re < 1000$

Max load = Max Stress × Area

$$B.F = B.s \times A$$

$$\Delta L \propto \frac{L}{r^2}$$

Capillary:

$$h = \frac{2\sigma \cos\theta}{r \rho g}$$

Surface Tension (σ)

$$\sigma = F/L$$

$\sigma \to$ Liquid fallen with Temperature

Excess Pressure inside a liquid drop:

$$P = \frac{2\sigma}{R}$$

Excess Pressure inside soap Bubble

$$P = \frac{4\sigma}{R}$$

Excess Pressure inside Air Bubble:

$$P = \frac{2\sigma}{R}$$

M.P.F

Chapter No - 10

THERMAL PROPERTIES OF MATTER:

$$\frac{°F - 32}{180} = \frac{°C}{100}$$

Absolute Zero Temperature

$$T \to -273 \cdot 15°C$$

→ Three Types of Expansion

1. Linear Expansion:

$$\alpha = \frac{\Delta L}{L} \times \frac{1}{\Delta T}$$

$$\Delta L = L(1 + \alpha \Delta T)$$

2. Areal Expansion,

$$\alpha = \alpha_1 + \alpha_2$$

isotropic → $\alpha = \frac{\Delta A}{A} \times \frac{1}{\Delta T}$

3. Volume Expansion:

$$\alpha = \frac{\Delta V}{V} \times \frac{1}{\Delta T}$$

$$\Delta V = V \gamma \Delta T$$

T.P.M

$$\alpha = \frac{\gamma}{3}$$

Thermal Stress

$$= Y \cdot \alpha \Delta T$$

• **Coefficient of Superficial Expansion:**

$$\beta = \frac{\Delta S}{S \Delta T}$$

• **Temperature Coefficient of Resistance:**

$$\alpha = \frac{R - R_0}{R_0 \times T}$$

• **Coefficient of Linear Expansion:**

$$\alpha = \frac{\Delta L}{L \Delta T}$$

- Heat gained/Heat lost:

$$Q = mc\Delta T$$

- Latent heat:

$$Q = mC$$

** Newton's law of cooling:

Rate of loss of heat
∝ Temperature difference b/w the body & the surrounding

$$m\,s = k\left(\left(\frac{T_1-T_2}{2}\right) - T_0\right)$$

- Coefficient of Cubical Vol expansion:

$$\gamma = \frac{\Delta V}{V \Delta T}$$

- Rate of flow of Heat:

$$\frac{dQ}{dt} = -kA\frac{dT}{dx}$$

$$\frac{dQ}{dt} = -kA\frac{(T_H - T_C)}{L}$$

T.P.M

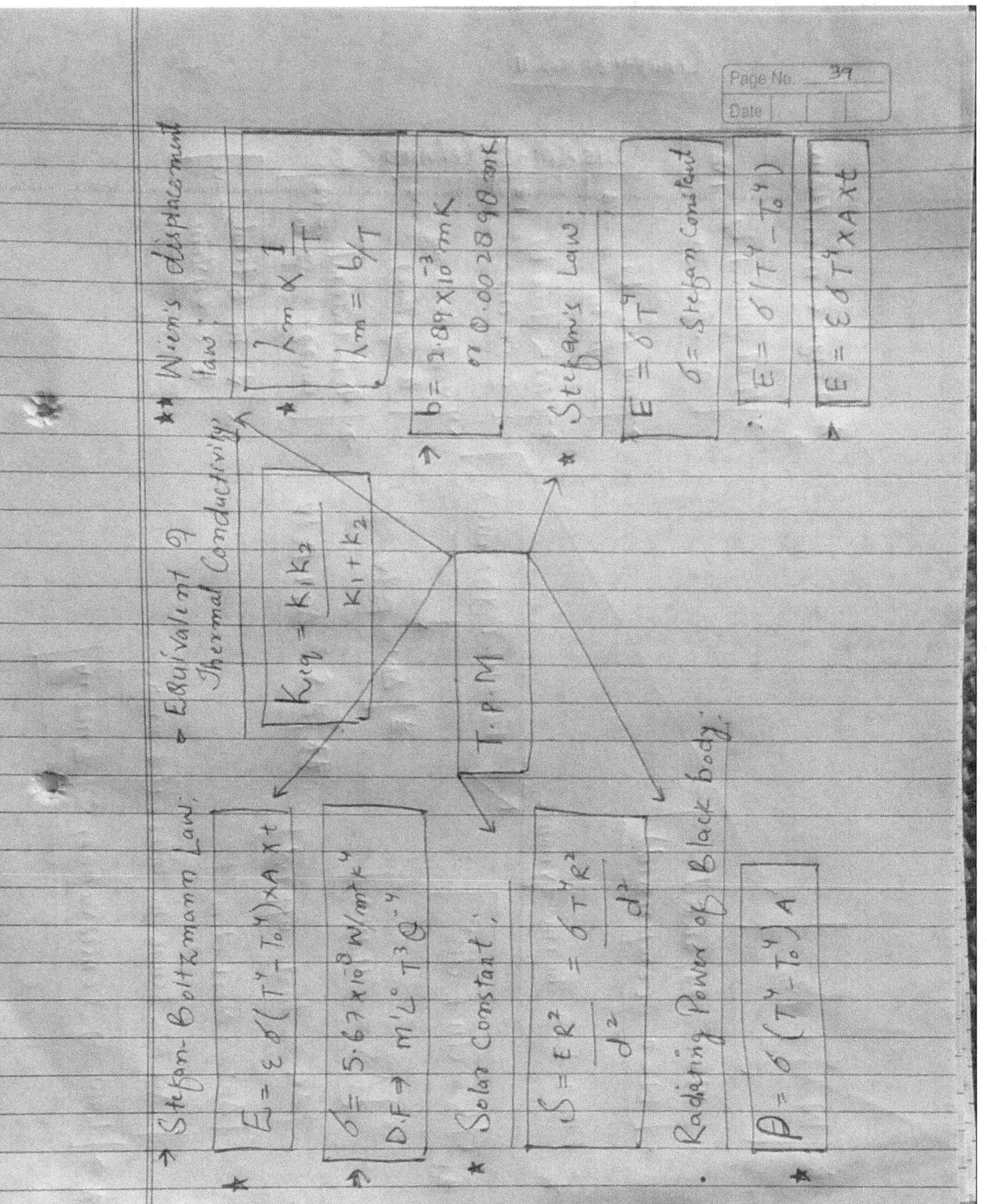

Chapter No. 11

THERMODYNAMICS

Sign Conventions:
- Heat gained by a system = +ve
- Heat loss by a system = −ve
- Work done by a system = +ve
- Work done on a system = −ve
- ↑ in internal energy of system = +ve
- ↓ in internal E of system = −ve

THERMODYNAMICS

★ $\Delta W = PdV$

★★ $\Delta Q = \Delta U + PdV$

★ $W = \int_{V_i}^{V_f} PdV$

- $W > 0$ (expanded)
- $W < 0$ (compressed)

• **First law Thermodynamic:**

★ $dQ = dU + dW$

or $\Delta Q = \Delta U + \Delta W$

- ΔQ = heat supplied
- ΔW = Work done
- ΔU = Change in internal Energy

★ → Law of Conservation of Energy

THERMODYNAMICS

• State Function:
→ Depends only upon initial + Final state of the System.

Ex: P, V, T, U, H, S.

•2 Path Function:
→ Depends upon the Path + initial + Final state of the System.

Ex: Work, Energy/heat

• Extensive properties:
→ Depends on the Quantity of Matter (Size and mass)

Ex: V, M, G, E, U, F, etc.

•2 Intensive Properties:
→ Independent of matter (Size and mass)

Ex: $P, T, PH, d, V_m, T_f, T_b, d$

★ Ratio of Two Extensive Property indicates the Intensive property.

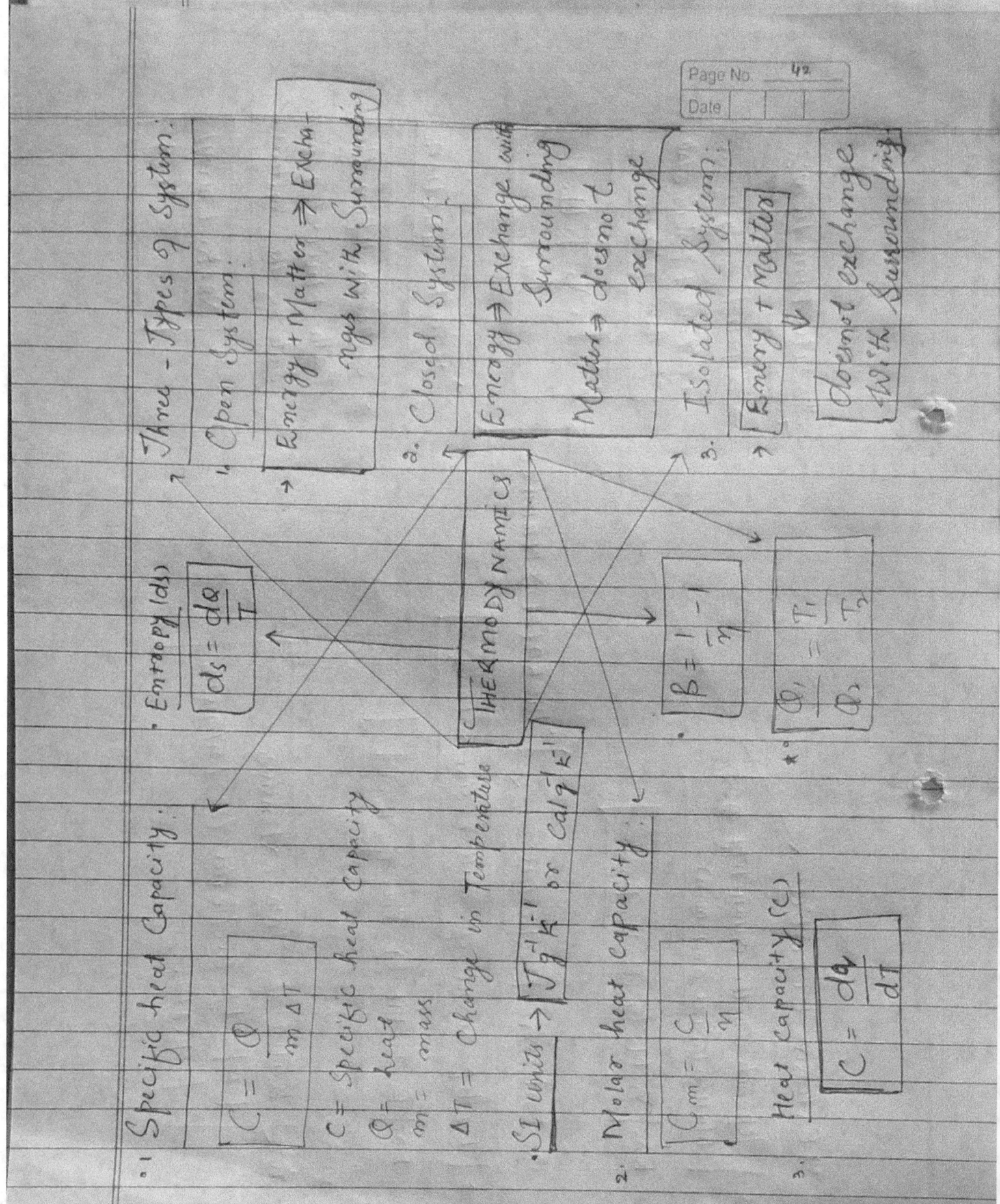

THERMODYNAMICS

- Three-Types of System:

1. **Open System:**
   → Energy + Matter ⇒ Exchanges with Surrounding

2. **Closed System:**
   → Energy ⇒ Exchange with Surrounding
   Matter ⇒ does not exchange

3. **Isolated System:**
   → Energy + Matter
   ↓
   does not exchange with Surrounding

- Entropy (ds)
  $$ds = \frac{dQ}{T}$$

- $\beta = \frac{1}{\eta} - 1$

- $\frac{Q_1}{Q_2} = \frac{T_1}{T_2}$

1. **Specific heat Capacity:**
   $$C = \frac{Q}{m\,\Delta T}$$
   C = Specific heat capacity
   Q = heat
   m = mass
   ΔT = Change in Temperature
   - SI units → $J\,g^{-1}\,K^{-1}$ or $Cal\,g^{-1}\,K^{-1}$

2. **Molar heat Capacity**
   $$C_m = \frac{C}{\eta}$$

3. **Heat capacity (C)**
   $$C = \frac{dq}{dT}$$

Types of Thermodynamics Process:

1. Isothermal:

* $\boxed{\Delta T = 0, \; \Delta U = 0}$

* $\boxed{W_{iso} = 2.303 \log \dfrac{P_1}{P_2}}$

* $\boxed{W_{iso} = 2.303 \log \dfrac{V_2}{V_1}}$

2. Isobaric Process:

* $\boxed{\Delta P = 0}$

$\boxed{W = P\Delta T}$

$\boxed{W = nRT}$

THERMODYNAMICS

- Polyatomic → $\gamma = 1.33$
- diatomic → $\gamma = 1.4$
- monoatomic → $\gamma = 1.66$

3. Adiabatic Process: Heat is neither absorbed nor released by a system

* $\boxed{\Delta Q = 0}$

* $\boxed{P_1 V_1^\gamma = P_2 V_2^\gamma}$

* $\boxed{T_1 V_1^{\gamma-1} = T_2 V_2^{\gamma-1}}$

* $\boxed{W_{ad} = \dfrac{P_2 V_2 - P_1 V_1}{\gamma - 1}}$

* $\boxed{W_{ad} = \dfrac{nR\Delta T}{\gamma - 1}}$

4. Isochoric Process:

* $\boxed{\Delta V = 0}$

$\boxed{W = -P\Delta V}$

$\boxed{W = 0}$

Atomicity of Gas	F	Cv	Cp	γ
Monoatomic	3	$\frac{3R}{2}$	$\frac{5R}{2}$	$\frac{5}{3} = 1.66$
Diatomic	5	$\frac{5R}{2}$	$\frac{7R}{2}$	$\frac{7}{5} = 1.4$
Polyatomic	6	$\frac{6R}{2} = 3R$	$\frac{8R}{2} = 4R$	$\frac{4}{3} = 1.33$

→ Change in Internal Energy (u)

$$\Delta u = \frac{MFR(dT)}{2} = MCvdT$$

$$Cv = \frac{FR}{2}$$

$$\Delta u = \frac{PV}{\gamma - 1}$$

$$\boxed{F = Freedom}$$

THERMODYNAMICS

1. Efficiency of heat engine:

$$\eta = \frac{W}{Q_1} = \frac{T_1 - T_2}{T_1} = \frac{Q_1 - Q_2}{Q_1}$$

2. Coefficient of Performance of a refrigerator:

$$\beta = \frac{Q_2}{W} = \frac{Q_2}{Q_1 - Q_2}$$

$$\beta = \frac{T_2}{T_1 - T_2}$$

3. Efficiency of Carnot's engine:

$$\eta = 1 - \frac{Q_2}{Q_1} = 1 - \frac{T_2}{T_1}$$

→ Total energy of each molecule:

$$E_T = \frac{FKT}{2}$$

Chapter No- 12
KINETIC THEORY

- Equation of state for ideal gas:

$$PV = nRT$$

$$\frac{P}{\rho} = \frac{RT}{M_w} = \frac{KT}{m}$$

- Van der Waals Equation

$$\left(P + \frac{an^2}{V^2}\right)(V - nb) = nRT$$

- Ratio of densities:

$$\frac{\rho_1}{\rho_2} = \left(\frac{P_1}{P_2}\right)\left(\frac{M_1}{M_2}\right)$$

1. **Boyle's Law:** [At Constant Temperature]

$$V \propto \frac{1}{P}$$

$$P_1 V_1 = P_2 V_2$$

2. **Charle's law:** [At Constant Pressure]

$$V \propto T$$

$$\frac{V_1}{V_2} = \frac{T_1}{T_2}$$

3. **Gay Lussac's law:** [At Constant Volume]

$$P \propto T$$

$$\frac{P_1}{P_2} = \frac{T_1}{T_2}$$

* Pressure exerted by a gas

$$P = \frac{1}{3} \frac{m}{V} v_{rms}^2 = \frac{1}{3} \rho v_{rms}^2$$

* Energy Related with each degree of freedom = $\frac{1}{2} kT$

★ $U = \frac{F}{2} kT$

★ $E_T = \frac{1}{2} m v_{rms}^2 = \frac{3}{2} PV$

★ $E_V = \frac{1}{2} \rho v_{rms}^2$

★ $\frac{v_1^2}{v_2^2} = \frac{T_1}{T_2}$

* Rms Speed of molecules:

$$V_{rms} = \sqrt{\frac{3P}{\rho}} = \sqrt{\frac{3RT}{Mm}} = \sqrt{\frac{3kT}{m}}$$

$$= 1.73 \sqrt{\frac{kT}{m}}$$

2. Average Speed or Mean Speed:

$$V_{av} = \sqrt{\frac{8P}{\pi\rho}} = \sqrt{\frac{8RT}{\pi Mm}} = \sqrt{\frac{8kT}{\pi m}} = 1.59\sqrt{\frac{kT}{m}}$$

3. Most Probable Speed

$$V_{mp} = \sqrt{\frac{2P}{\rho}} = \sqrt{\frac{2RT}{Mm}} = \sqrt{\frac{2kT}{m}} = 1.41\sqrt{\frac{kT}{m}}$$

$1 k_B T$

$E = \frac{3}{2} k_B T$

Chapter - No. 13

Oscillations:

[Oscillations]

- Force displacement Relation in SHM

  $$F = -ky$$

  $$k = m\omega^2$$

- General Equation of Displacement in SHM

  $$y = A\sin(\omega t + \phi)$$

  or

  $$y = A\cos(\omega t + \phi)$$

  Angular frequency

  $$\omega = \frac{2\pi}{T} = 2\pi\nu$$

1. Velocity in SHM

   $$V = \frac{dy}{dt} = \omega A \cos(\omega t + \phi)$$

   $$V = \omega\sqrt{A^2 - y^2}$$

   $$V_{max} = \omega A$$

2. Acceleration:

   $$a = \frac{dv}{dt} = -\omega^2 y$$

   $$a_{max} = \omega^2 A$$

3. Time:

   $$T = \frac{2\pi}{\omega}$$

   $$T = 2\pi\sqrt{\frac{y}{a}} = 2\pi\sqrt{\frac{m}{k}}$$

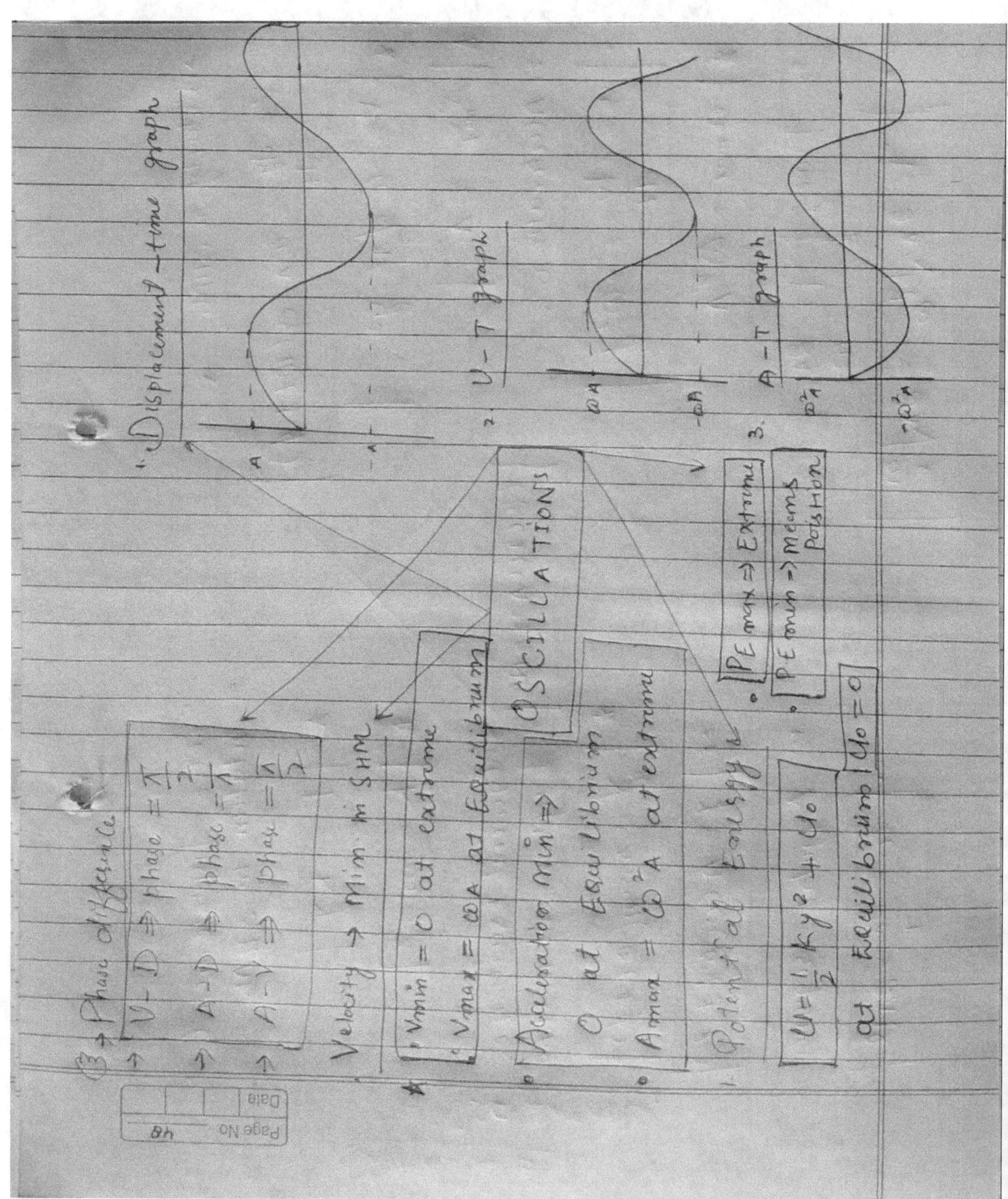

OSCILLATIONS

1. Displacement – time graph
2. V – T graph
3. A – T graph

③ → Phase difference
- V – D ⇒ Phase = $\frac{\pi}{2}$
- A – D ⇒ Phase = π
- A – V ⇒ Phase = $\frac{\pi}{2}$

★ Velocity → Min. in SHM
- $V_{min} = 0$ at extreme
- $V_{max} = \omega A$ at Equilibrium

• Acceleration Min ⇒
- 0 at Equilibrium
- $A_{max} = \omega^2 A$ at extreme

• Potential Energy
- $PE_{max} \Rightarrow$ Extreme
- $PE_{min} \Rightarrow$ Mean Position

$$U = \frac{1}{2} ky^2 + U_0$$

at Equilibrium $U_0 = 0$

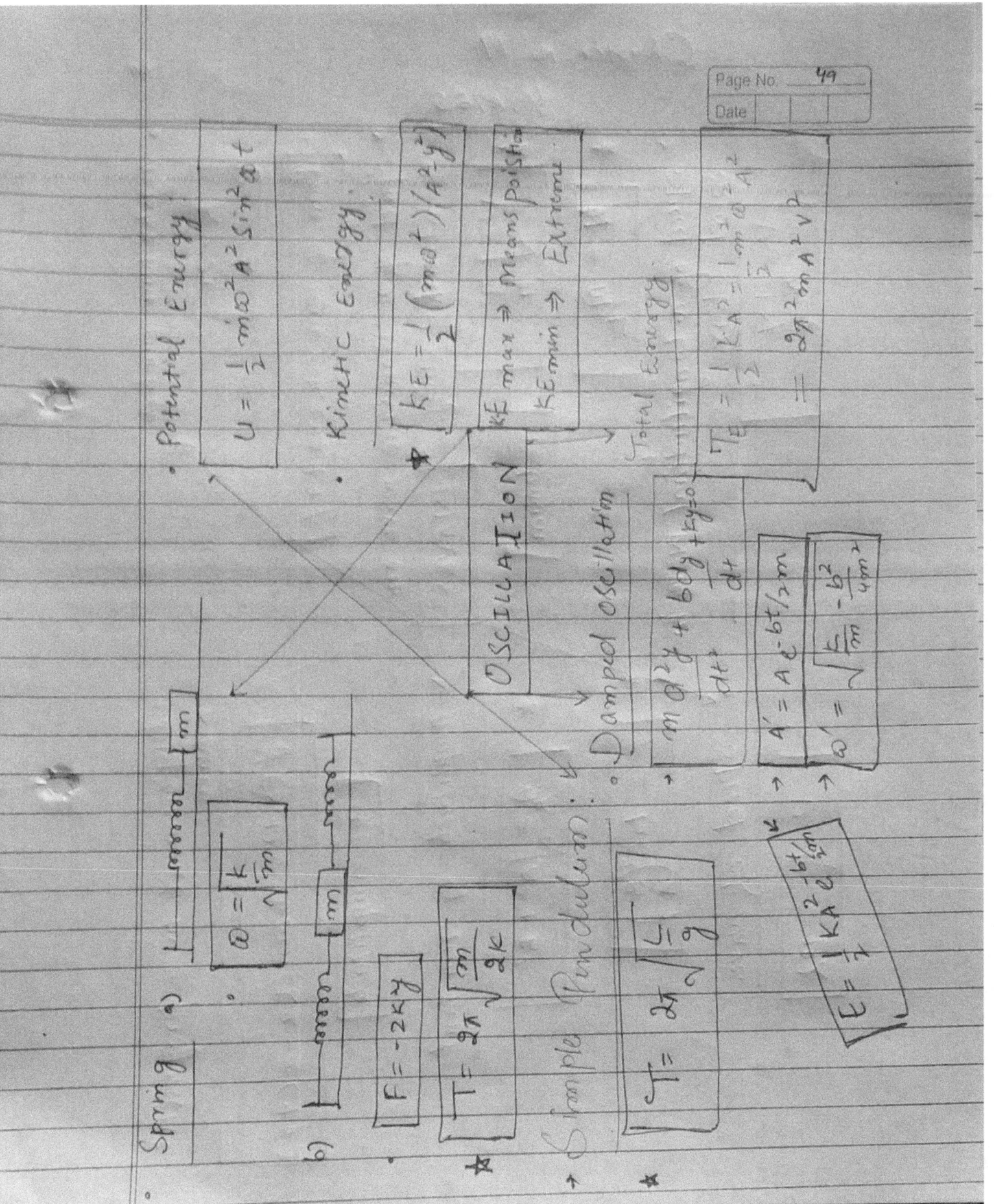

OSCILLATION

Spring

a) [spring with mass m]

$$\omega = \sqrt{\dfrac{k}{m}}$$

b) [spring with mass m]

$$F = -2ky$$

$$T = 2\pi \sqrt{\dfrac{m}{2k}}$$

Simple Pendulum:

$$T = 2\pi \sqrt{\dfrac{L}{g}}$$

- **Potential Energy**

$$U = \tfrac{1}{2} m\omega^2 A^2 \sin^2 \omega t$$

- **Kinetic Energy**

$$KE = \tfrac{1}{2}(m\omega^2)(A^2 - y^2)$$

★ $KE_{max} \Rightarrow$ Mean Position
  $KE_{min} \Rightarrow$ Extreme

- **Total Energy**

$$T.E = \tfrac{1}{2} kA^2 = \tfrac{1}{2} m\omega^2 A^2$$

$$= 2\pi^2 \nu^2 m A^2$$

- **Damped Oscillation**

$$m\dfrac{d^2y}{dt^2} + b\dfrac{dy}{dt} + ky = 0$$

→ $A' = A e^{-bt/2m}$

→ $\omega' = \sqrt{\dfrac{k}{m} - \dfrac{b^2}{4m^2}}$

★ $E = \tfrac{1}{2} kA^2 e^{-bt/m}$

Chapter No. 14
Waves

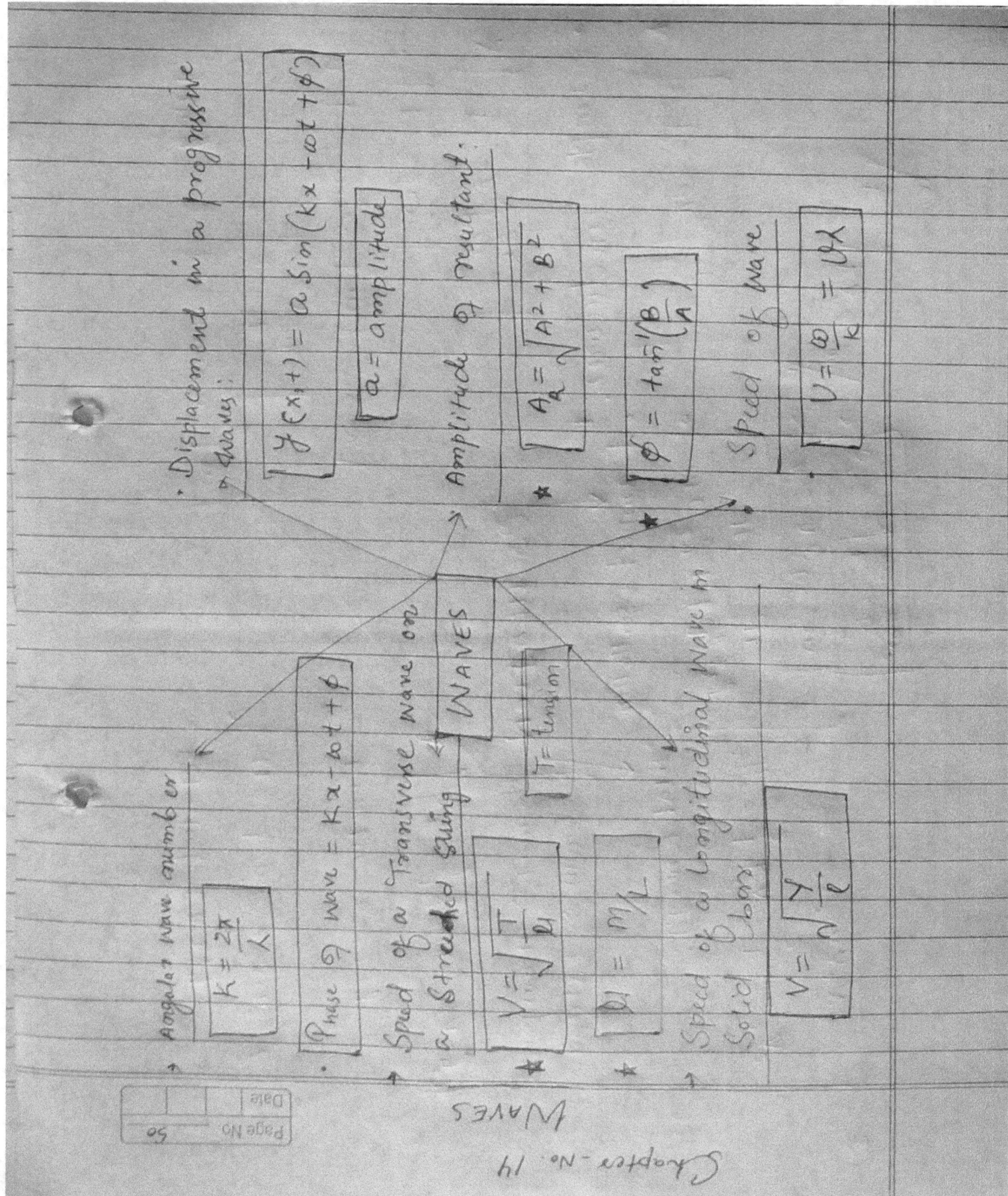

- Angular wave number
  $$k = \frac{2\pi}{\lambda}$$

- Phase of wave $= kx - \omega t + \phi$

- Speed of a Transverse wave on a Stretched String
  $$v = \sqrt{\frac{T}{\mu}}$$
  $$\mu = m/L$$
  T = tension

- Speed of a Longitudinal wave in Solid bars
  $$v = \sqrt{\frac{Y}{\rho}}$$

- Displacement in a progressive wave:
  $$y(x,t) = a \sin(kx - \omega t + \phi)$$
  a = amplitude

- Amplitude of resultant:
  $$A_R = \sqrt{A^2 + B^2}$$
  $$\phi = \tan^{-1}\left(\frac{B}{A}\right)$$

- Speed of wave
  $$v = \frac{\omega}{k} = \nu\lambda$$

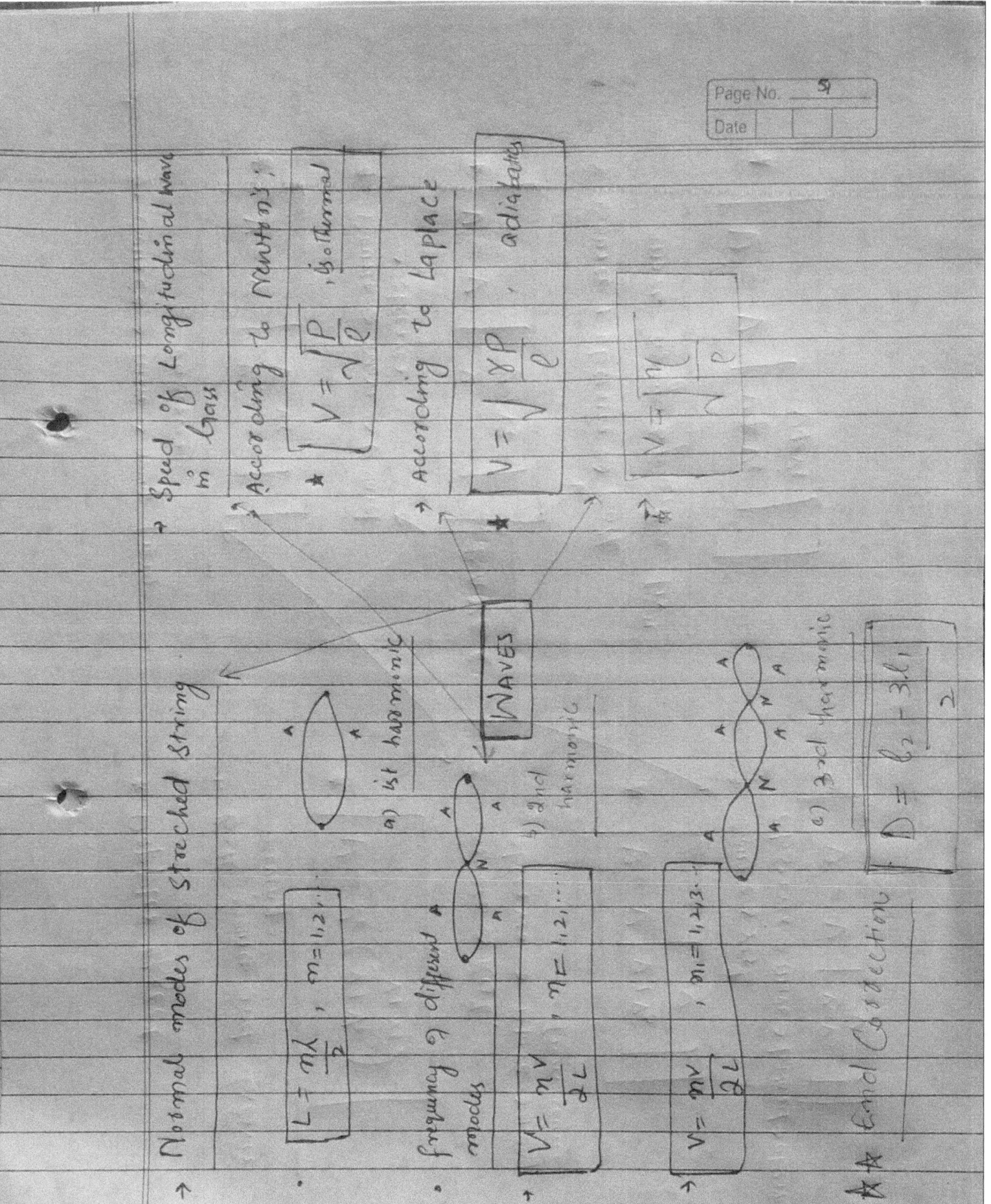

WAVES

Doppler Effect:

1. Both Source and observer moving:

$$V' = V_0 \left(\frac{V + V_0}{V + V_s} \right)$$

2. Source → Rest, observer ⇒ Move → away from Source

$$V' = V \left[\frac{V + V_0}{V} \right]$$

3. Source → rest, observer ⇒ Move → toward → observer

$$V' = V \left[\frac{V - V_0}{V} \right]$$

3. Source → Move → Away from observer:

$$V' = V \left[\frac{V}{V - V_s} \right]$$

4. Source ⇒ Move → toward → observer

$$V' = V \left[\frac{V}{V + V_s} \right]$$

5. Source + observer ⇒ Both move away opposite direction.

$$V' = V \left[\frac{V + V_s}{V + V_s} \right]$$

WAVES

- Source + observer → Both Move → toward each others.

  ★ $$U' = v\left[\dfrac{v+v_o}{v-v_s}\right]$$

- Fundamental frequency:

  ★ $$\vartheta = \dfrac{v}{2\ell} = \dfrac{1}{2\ell}\sqrt{\dfrac{T}{m}}$$

  ★ $$v = \sqrt{\dfrac{\gamma RT}{M}}$$

- Closed pipe at one end, only odd harmonics.

  ▷ Fundamental mode: $$\boxed{V_1 = \dfrac{V}{4\ell} = \vartheta}$$ (1st harmonic)

  ▷ Second mode: $$\boxed{V_2 = 3V}$$

  ▷ n^{th} mode: $$\boxed{V_n = (2n-1)V}$$

- Open pipe at both ends.
  Both odd + Even:

  ▷ Fundamental mode: $$\boxed{\vartheta_1 = \dfrac{v}{2\ell} = v'}$$

  ▷ Second mode: $$\boxed{V_2 = 2v'}$$

  ▷ n^{th} mode: $$\boxed{V_n' = nv'}$$

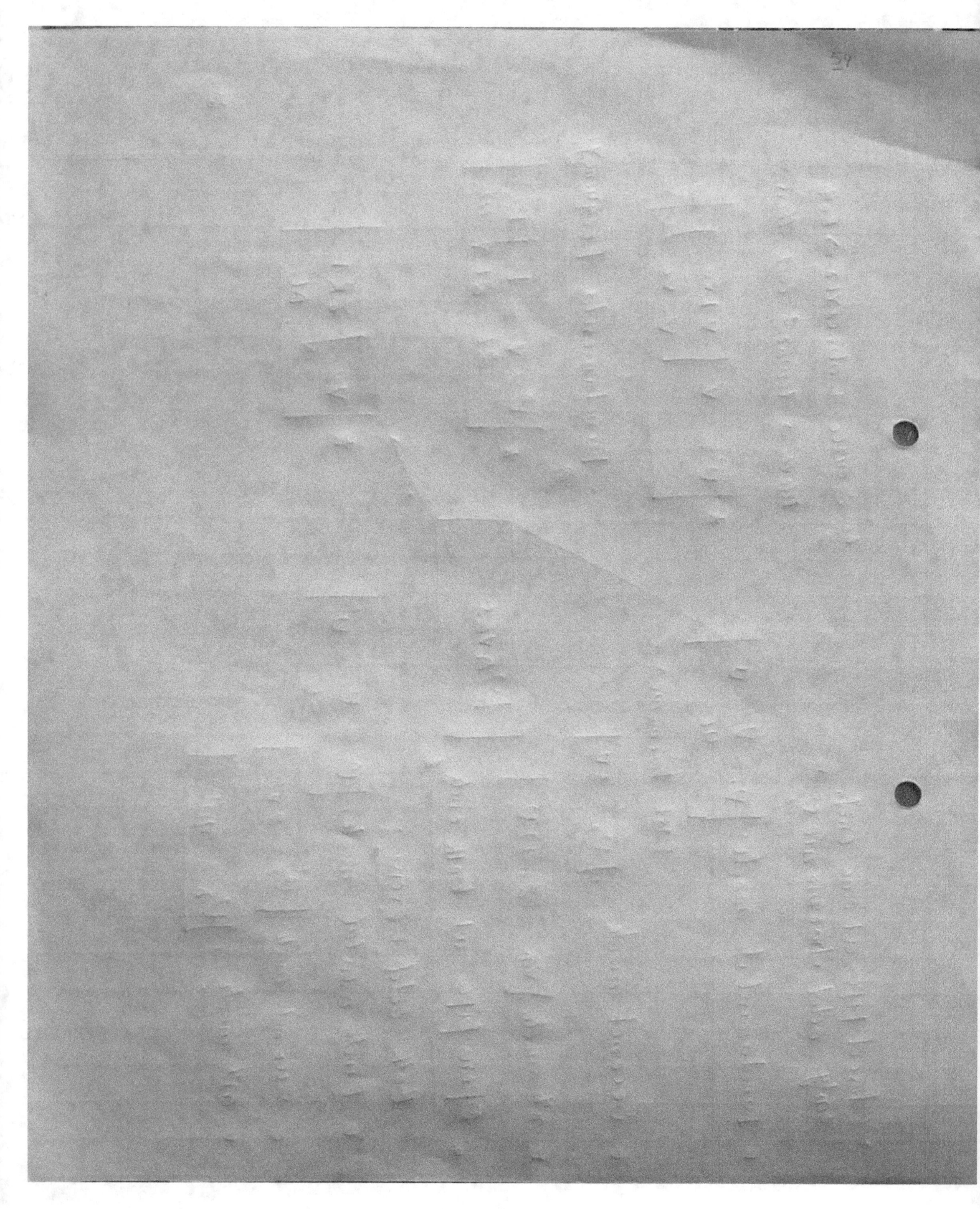

PBM

Quick Revision
Chapterwise
Formula's

BRAIN-MAP

PHYSICS
Vol. 2

Class-12th

Formula's extracted from each line of NCERT Textbooks.

Azhar Mahmood Mir

EDITION → 2022

[BRAIN]
↑
[P B T]
↙   ↘
[PHYSICS]   [MAP]

"ONCE YOUR MIND STRETCHES TO A NEW LEVEL IT NEVER BACK TO ITS ORIGINAL DIMENSION"

Contents

Chapter 1 — ELECTRIC CHARGES AND FIELDS.
Chapter 2 — ELECTROSTATIC POTENTIAL AND CAPCITANCE.
Chapter 3 — CURRENT ELECTRICITY.
Chapter 4 — MOVING CHARGES AND MAGNETISM.
Chapter 5 — MAGNETISM AND MATTER.
Chapter 6 — ELECTROMAGNETIC INDUCATION.
Chapter 7 — ALTERNATING CURRENT.
Chapter 8 — ELECTROMAGNETIC WAVES
Chapter 9 — RAY OPTICS AND OPTICAL INSTRUMENT
Chapter 10 — WAVE OPTICS
Chapter 11 — DUAL NATURE OF RADIATION AND MATTER.
Chapter 12 — ATOMS
Chapter 13 — NUCLEI.
Chapter 14 — SEMICONDUCTOR ELECTRONICS: MATERAIS, DEVICES AND SIMPLE CIRCUIT.
Chapter 15 — COMMUNICATION SYSTEMS.

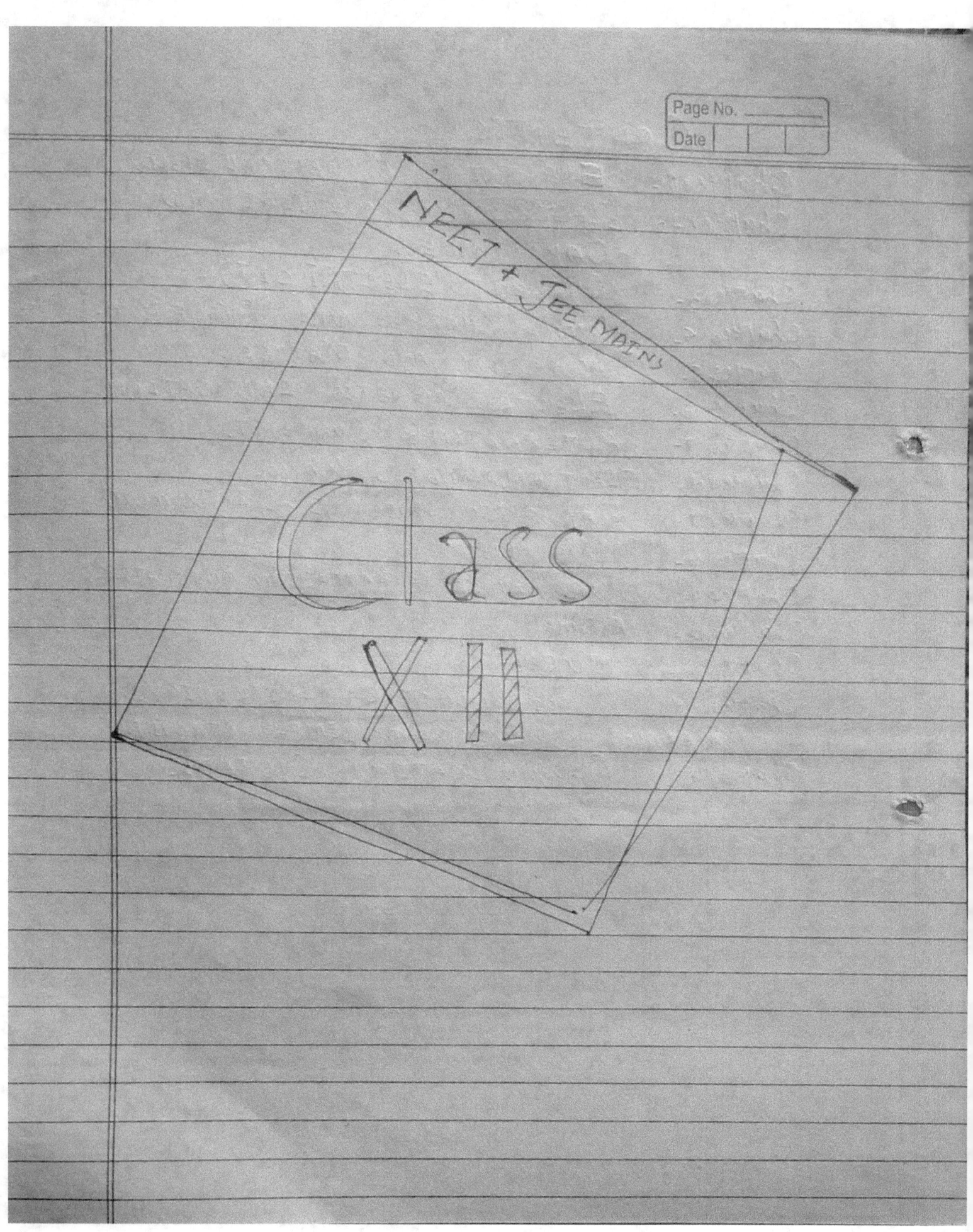

Chapter No-1 — Electric Charges and Fields

★ ELECTROSTATIC (ELECTRIC Charges & Fields)

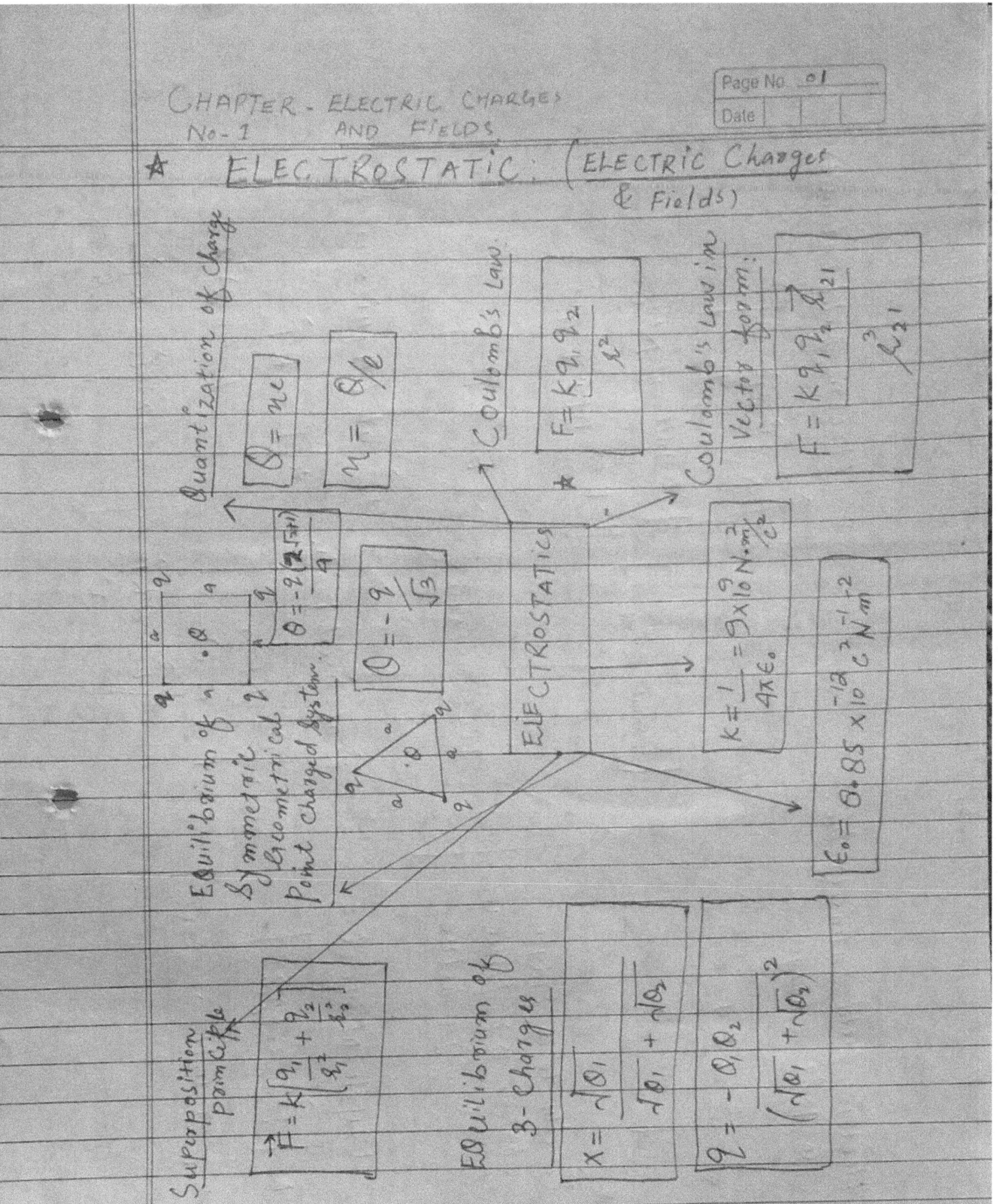

Quantization of charge
$$Q = ne$$
$$n = Q/e$$

Coulomb's Law
$$F = \frac{k q_1 q_2}{r^2}$$

Coulomb's law in Vector form:
$$\vec{F} = \frac{k q_1 q_2}{r^3} \vec{r}_{21}$$

ELECTROSTATICS
$$k = \frac{1}{4\pi\varepsilon_0} = 9 \times 10^9 \, N\cdot m^2/C^2$$
$$\varepsilon_0 = 8.85 \times 10^{-12} \, C^2 N^{-1} m^{-2}$$

Equilibrium of Symmetric 2 Geometrical Point Charged System

$$Q = -q\left(\frac{2\sqrt{2}+1}{4}\right)$$

$$Q = -\frac{q}{\sqrt{3}}$$

Superposition Principle
$$\vec{F} = k\left(\frac{q_1}{r_1^2} + \frac{q_2}{r_2^2}\right)$$

Equilibrium of 3-charges
$$x = \frac{\sqrt{Q_1}}{\sqrt{Q_1} + \sqrt{Q_2}}$$

$$q = -\frac{Q_1 Q_2}{(\sqrt{Q_1} + \sqrt{Q_2})^2}$$

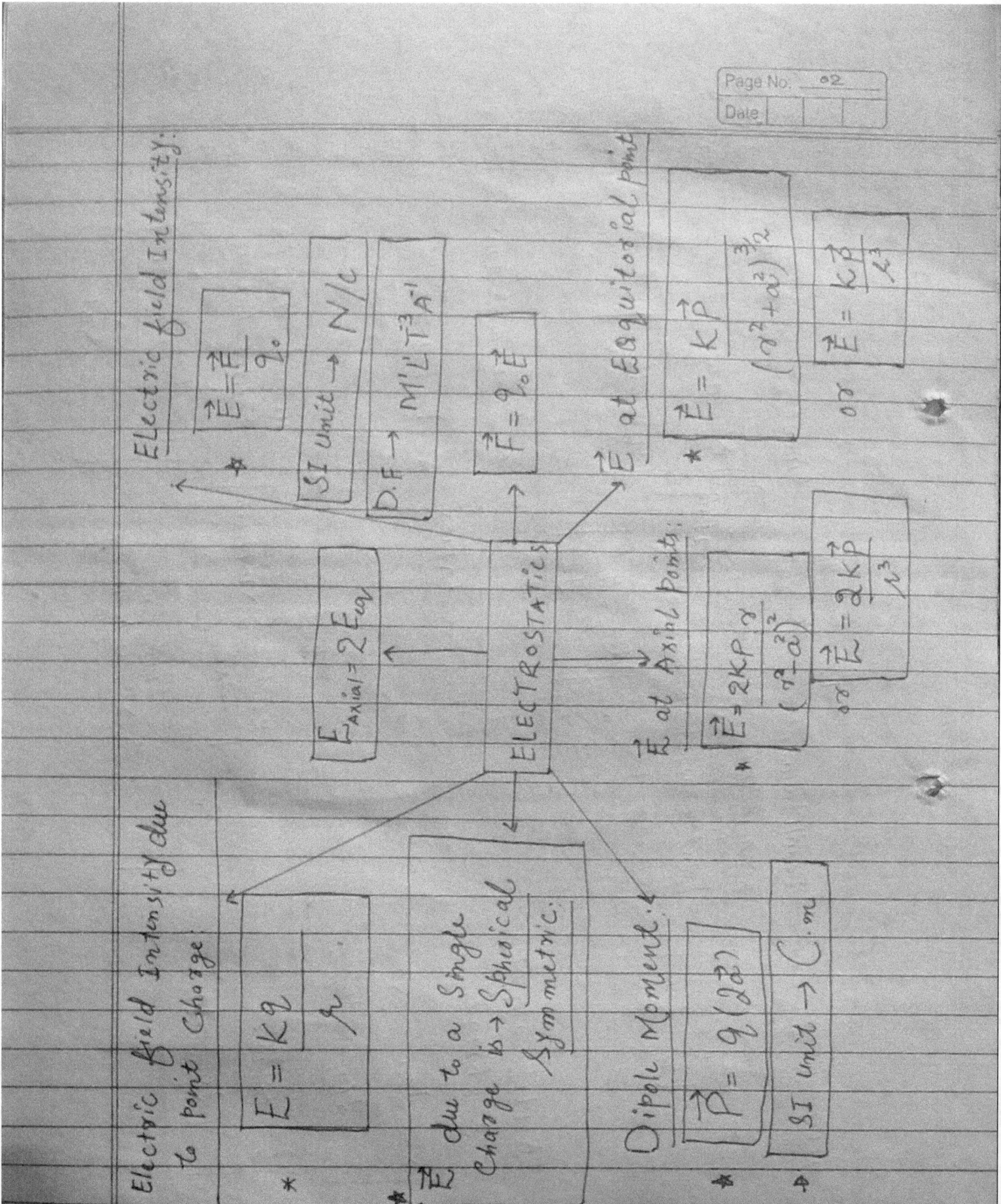

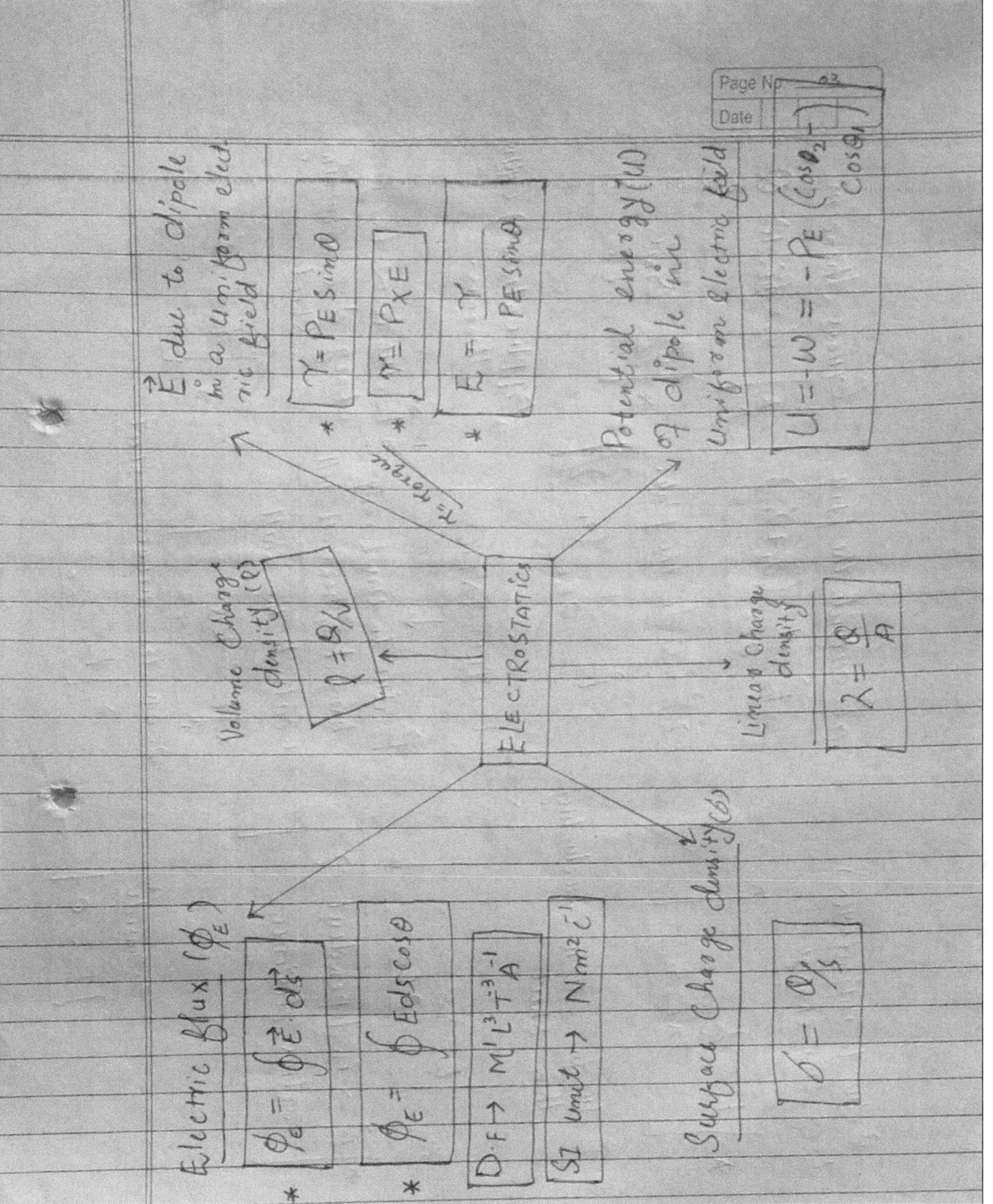

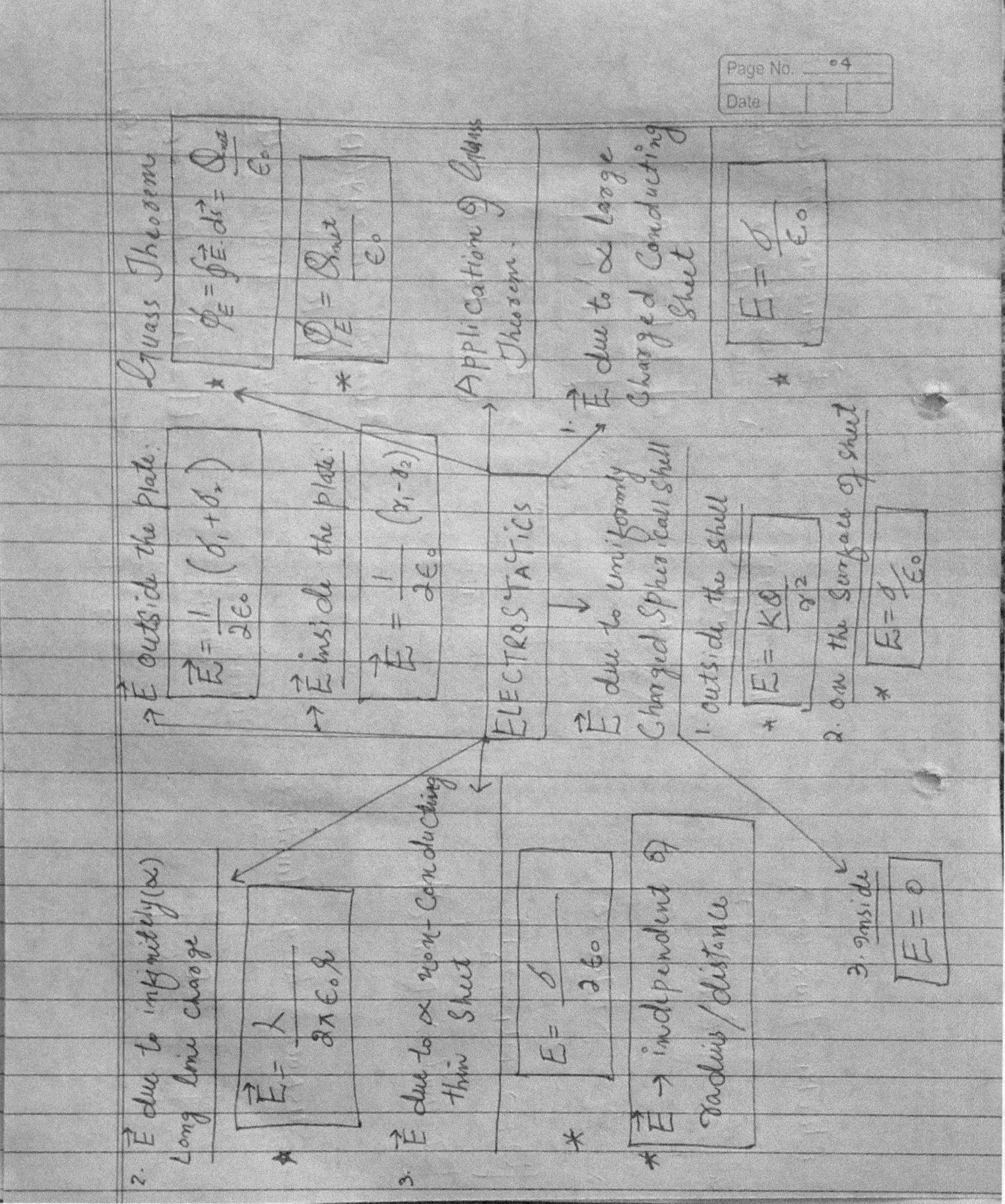

ELECTROSTATICS

Gauss Theorem

$$\phi_E = \oint \vec{E} \cdot d\vec{s} = \frac{Q_{in}}{\epsilon_0}$$

* $\boxed{\phi_E = \frac{Q_{in}}{\epsilon_0}}$

Application of Gauss Theorem

1. → $\vec{E}$ due to ∞ large Charged Conducting Sheet

* $\boxed{E = \dfrac{\sigma}{\epsilon_0}}$

→ $\vec{E}$ outside the plate:
$$\vec{E_1} = \frac{1}{2\epsilon_0}(\sigma_1 + \sigma_2)$$

→ $\vec{E}$ inside the plate:
$$\vec{E_2} = \frac{1}{2\epsilon_0}(\sigma_1 - \sigma_2)$$

2. → $\vec{E}$ due to infinitely(∞) long line charge

* $\boxed{\vec{E} = \dfrac{\lambda}{2\pi\epsilon_0 R}}$

3. → $\vec{E}$ due to ∞ non-Conducting thin Sheet

$\boxed{E = \dfrac{\sigma}{2\epsilon_0}}$

* $\vec{E}$ → independent of radius/distance

→ $\vec{E}$ due to uniformly Charged Spherical Shell

1. outside the shell
* $\boxed{E = \dfrac{KQ}{r^2}}$

2. on the Surface of shell
* $\boxed{E = \sigma/\epsilon_0}$

3. inside
$\boxed{E = 0}$

Chapter No-2: Electrostatic Potential and Capacitance

★ Electric Potential:

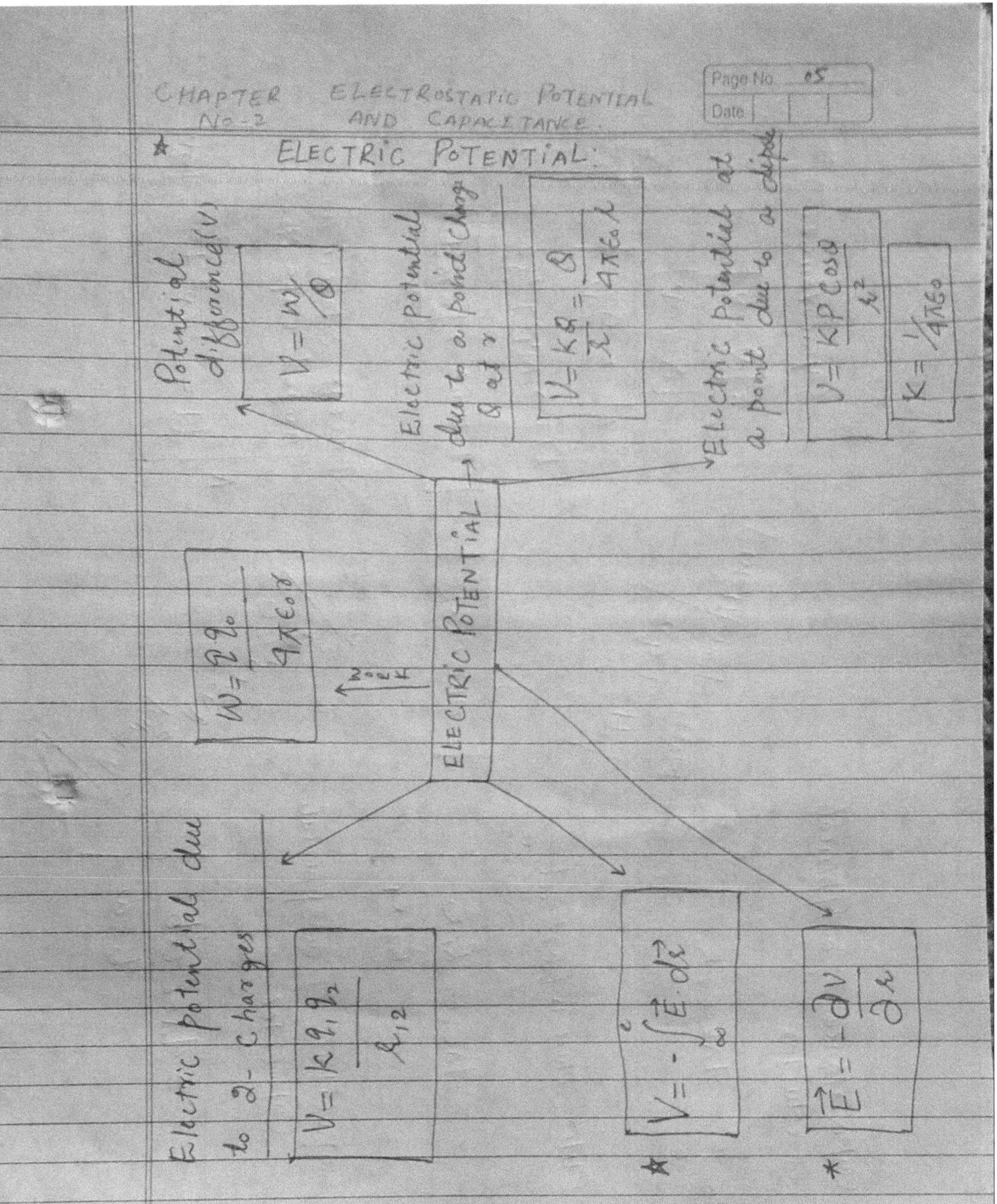

Potential difference (V)
$$V = W/Q$$

Electric Potential due to a point charge Q at r
$$V = \frac{KQ}{r} = \frac{Q}{4\pi\varepsilon_0 r}$$

Electric Potential at a point due to a dipole
$$V = \frac{KP\cos\theta}{r^2}$$
$$K = \frac{1}{4\pi\varepsilon_0}$$

$$W = \frac{q\, q_0}{4\pi\varepsilon_0 d}$$

Electric Potential due to 2-charges
$$V = \frac{K q_1 q_2}{r_{12}}$$

$$V = -\int_\infty^r \vec{E}\cdot d\vec{r}$$

$$\vec{E} = -\frac{\partial V}{\partial r}$$

CAPACITOR

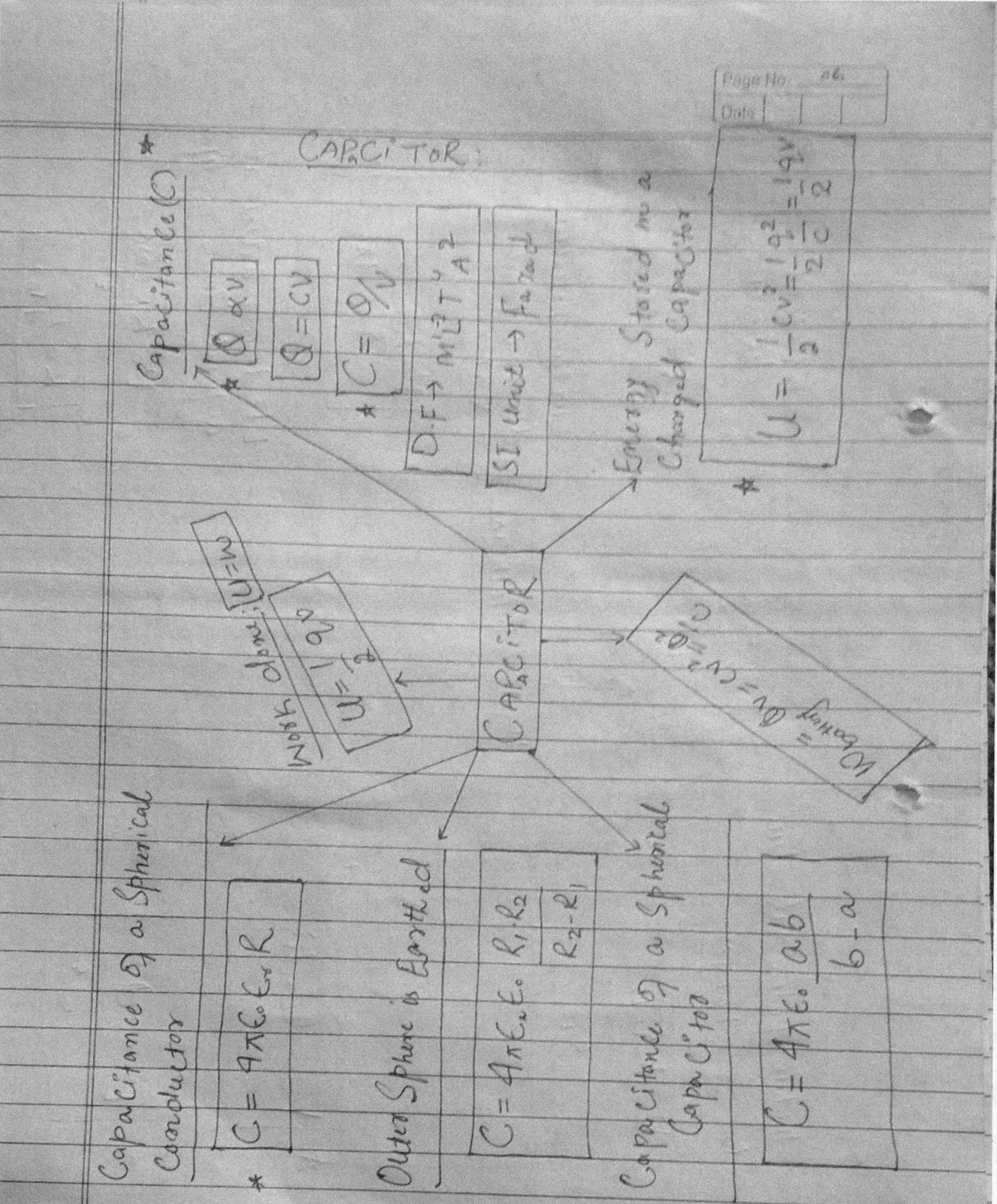

Capacitance (C)

- $Q \propto V$
- $Q = CV$
- $C = Q/V$

* D.F → $M^{-1}L^{-2}T^4A^2$
* SI Unit → Farad

Energy Stored in a Charged Capacitor

$$U = \frac{1}{2}CV^2 = \frac{1}{2}\frac{Q^2}{C} = \frac{1}{2}QV$$

Work done by Battery: $W = QV = CV^2$

Work done on $U = W$: $U = \frac{1}{2}Q^2$

Capacitance of a Spherical Conductor

* $C = 4\pi\varepsilon_0 R$

Outer Sphere is Earthed

$C = 4\pi\varepsilon_r\varepsilon_0 \dfrac{R_1 R_2}{R_2 - R_1}$

Capacitance of a Spherical Capacitor

$C = 4\pi\varepsilon_0 \dfrac{ab}{b-a}$

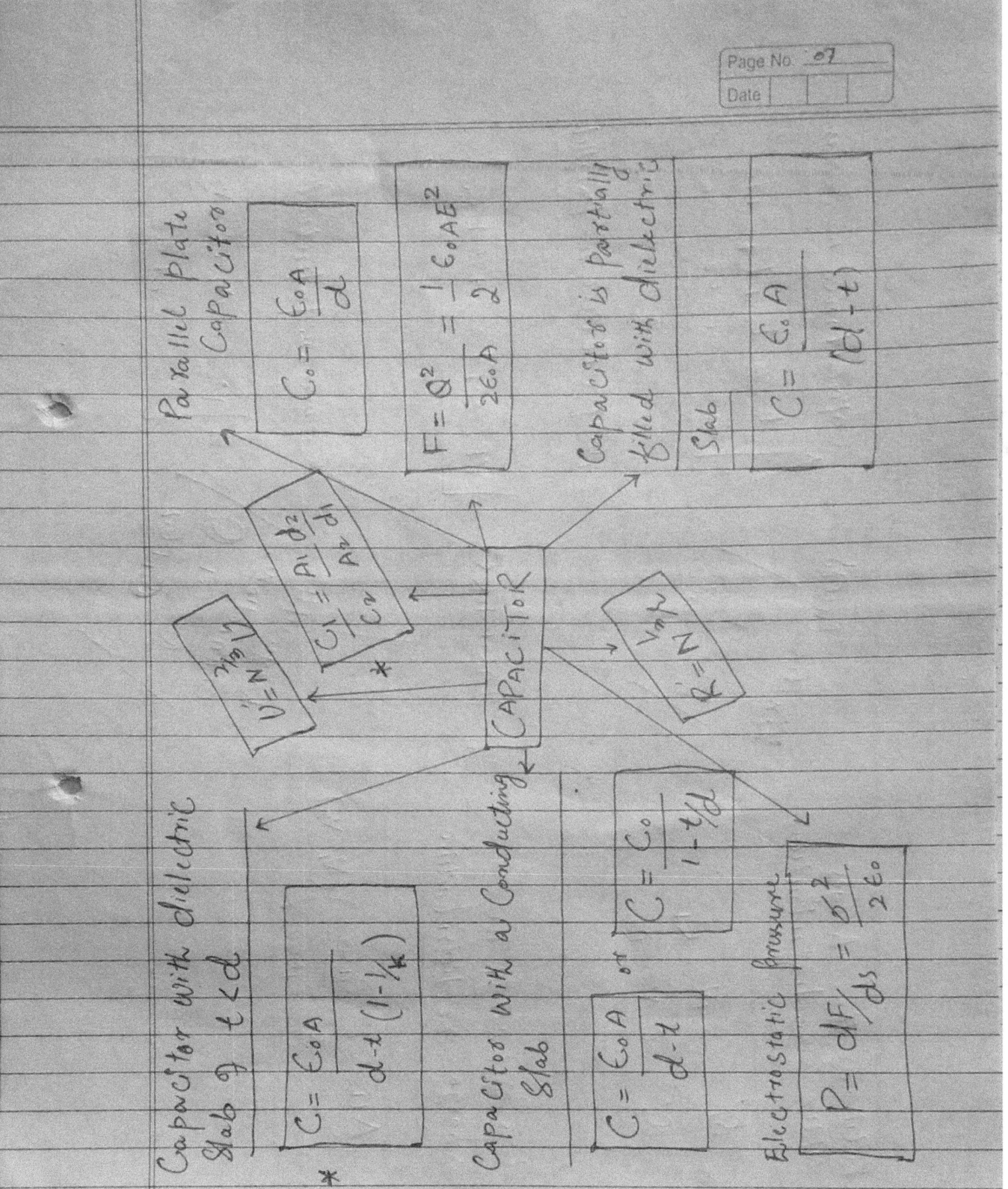

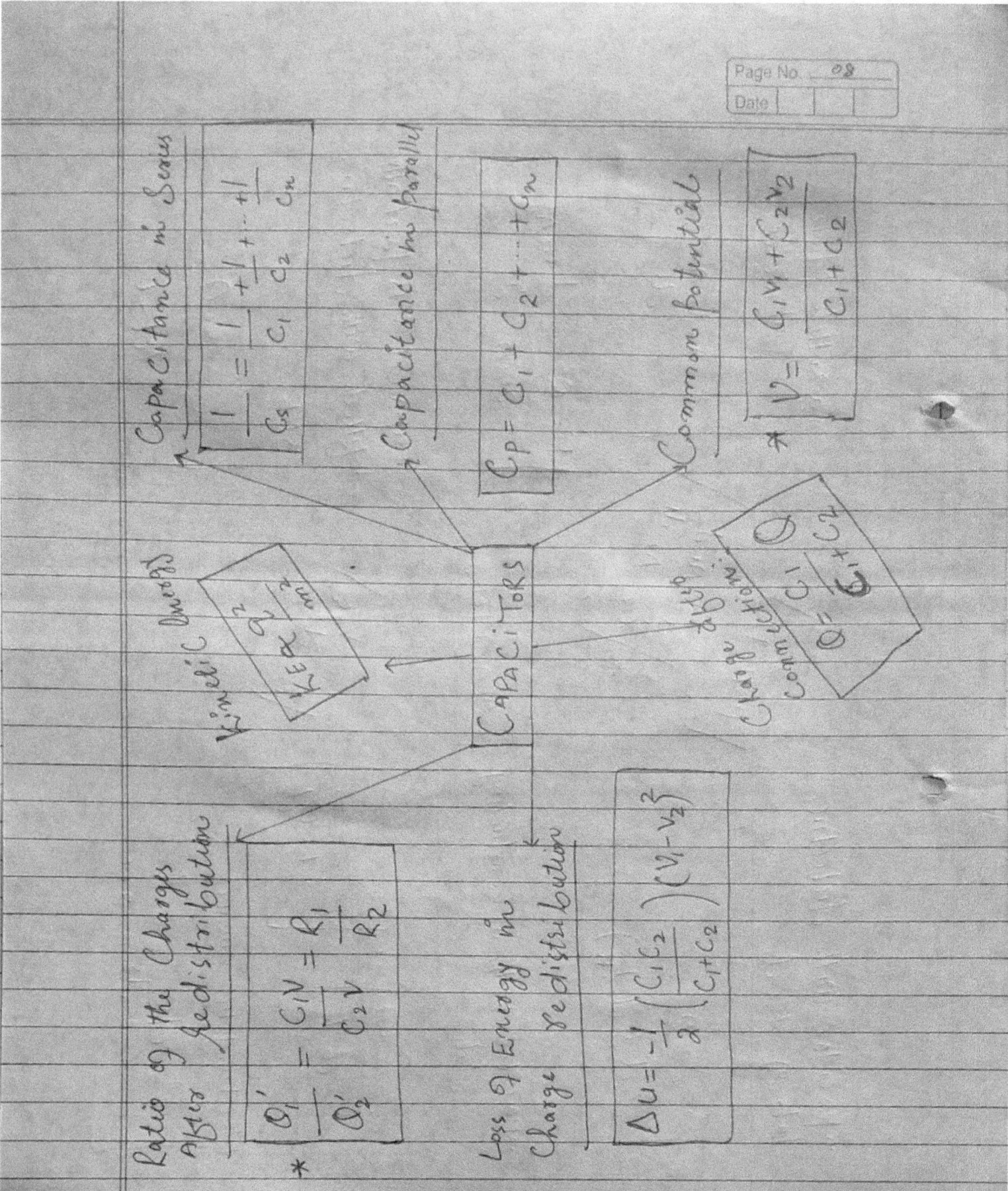

CURRENT ELECTRICITY (CE)

Current
$$I = \frac{Q}{t} = \frac{ne}{t}$$
$$I = \frac{dq}{dt}$$

SI unit → Ampere
D.F → $M^0 L^0 T^0 A^1$

Current density ($\vec{J}$)
$$J = I/A = ne\, v_d$$
$$I = \int \vec{J} \cdot d\vec{A}$$
SI unit → $A\,m^{-2}$
D.F → $M^0 L^{-2} T^0 A^1$

Relaxation Time (τ)
$$\tau = \frac{\tau_1 + \tau_2 + \cdots + \tau_n}{n}$$

Charge Flow = Average $i \times t$

Ohm's Law
$$V = IR$$
$$R = V/I$$
$$R = \frac{m\ell}{An e^2 \tau}$$

Drift Velocity
$$\vec{v_d} = -\frac{e\vec{E}\tau}{m}$$

Relation b/w Current and Drift Velocity
$$I = neA v_d$$

Relation b/w $\vec{J}, \vec{\sigma}, \vec{E}$
$$\vec{J} = \sigma \vec{E}$$
↓
Microscopic Ohm's law

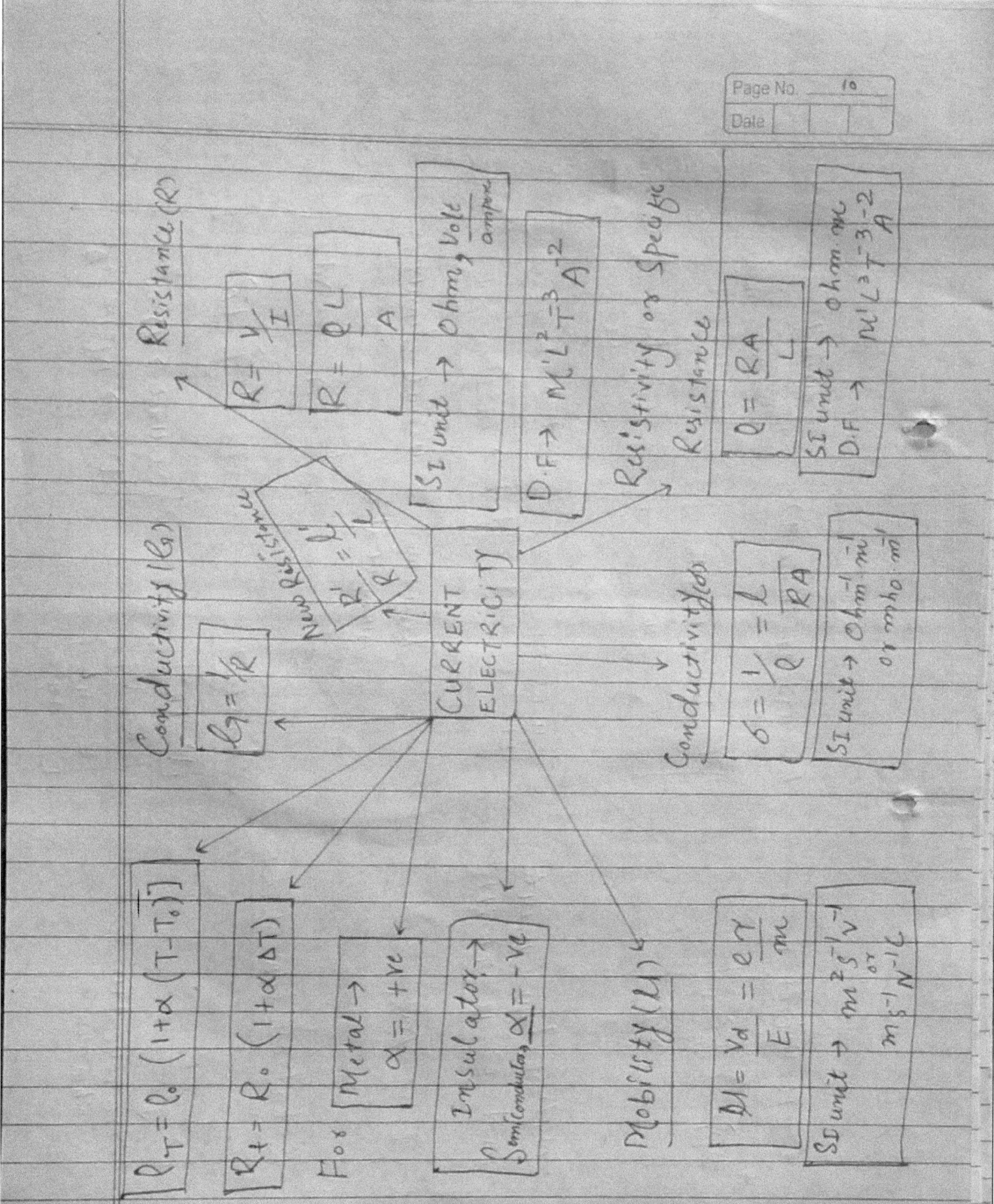

Colour Coding of Carbon resistor

Multiplier	Number	Colour
10^0	0	B → Black
10^1	1	B → Brown
10^2	2	R → Red
10^3	3	O → Orange
10^4	4	Y → Yellow
10^5	5	Great → Green
10^6	6	Britain → Blue
10^7	7	Very → Violet
10^8	8	Good → Gray
10^9	9	Nig → White

Tolerance	
20%	Nickel → Nickel
10%	Silver → Silver
5%	Gold → Gold

→ Strip 1st : Indicate the colour number of 'n' in resistance ohm.

→ Strip 2nd : Indicate the colour number of 'n' in resistance ohm.

→ Strip 3rd : Indicate Multiplier

→ Strip 4th : Indicate Tolerance

Example:

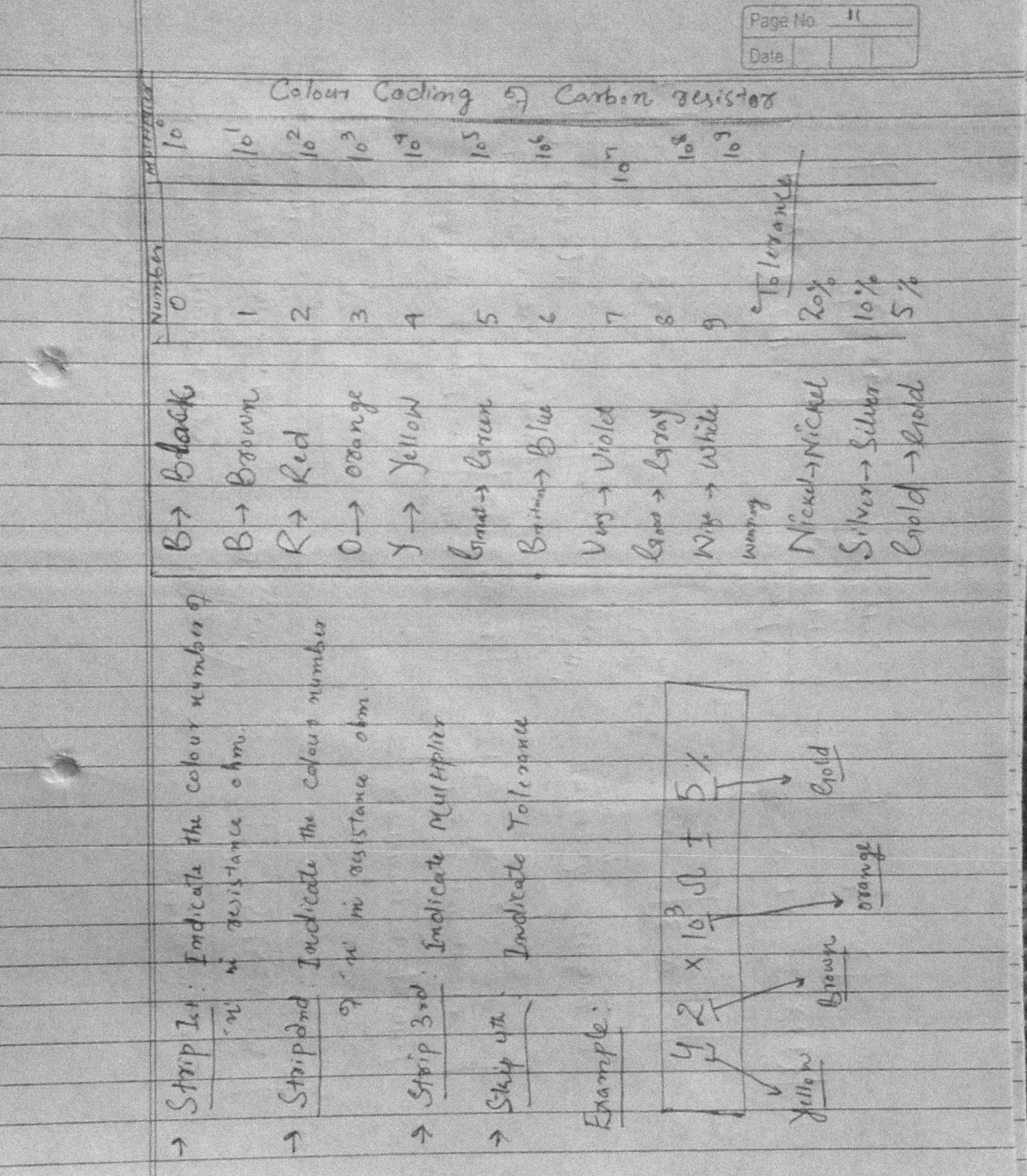

$4\ 2\ \times 10^3\ \Omega \pm 5\%$

Yellow Brown Orange Gold

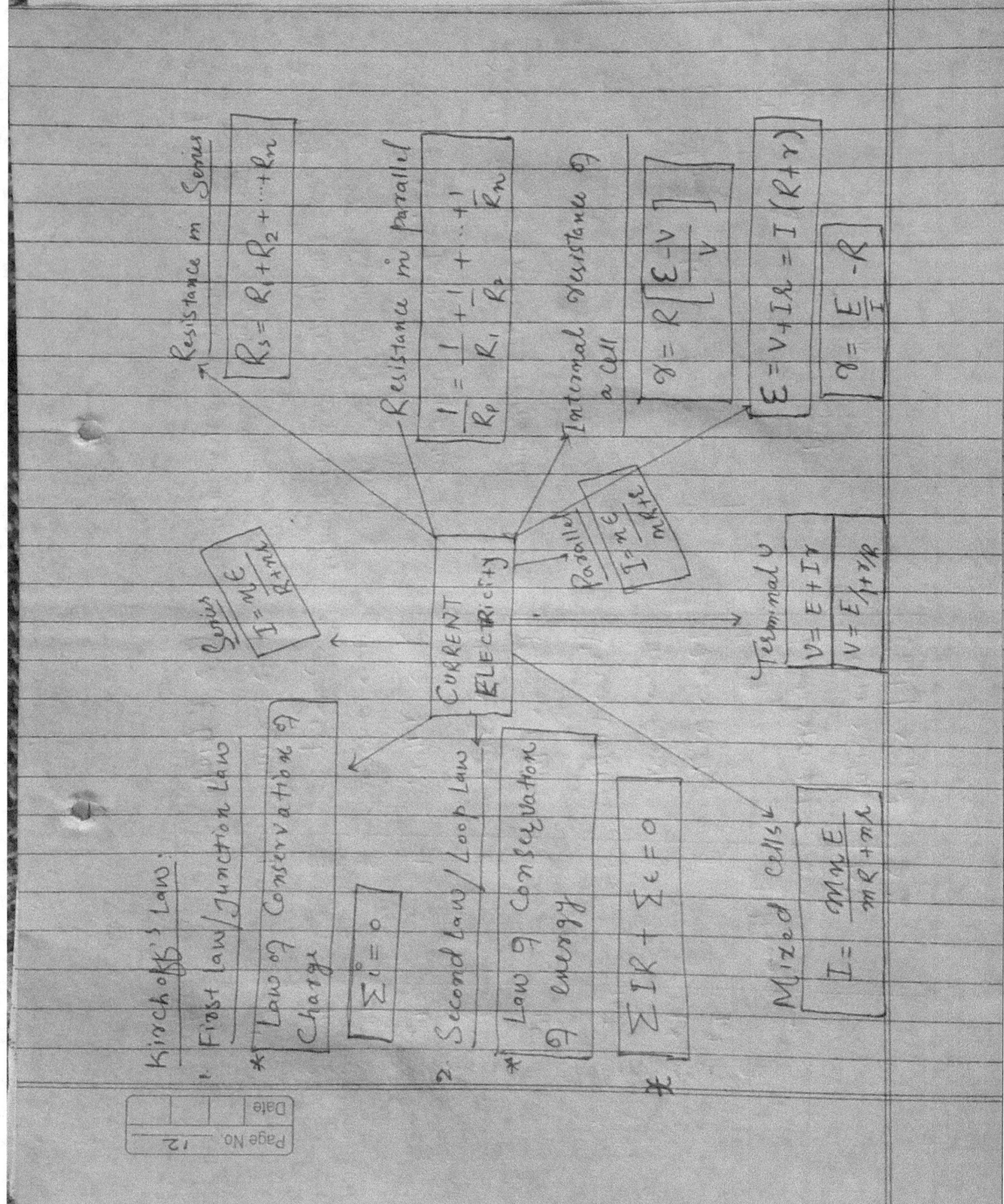

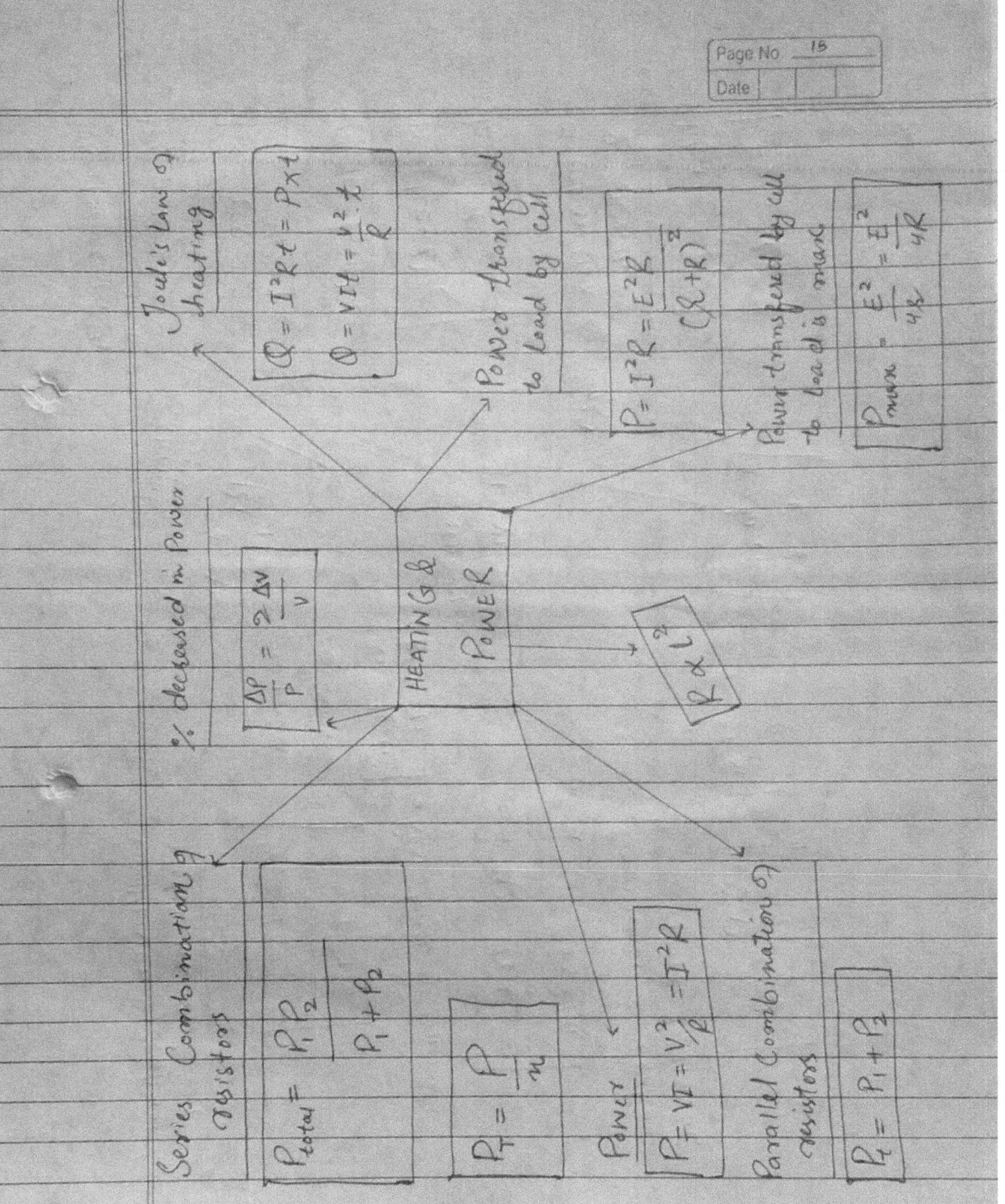

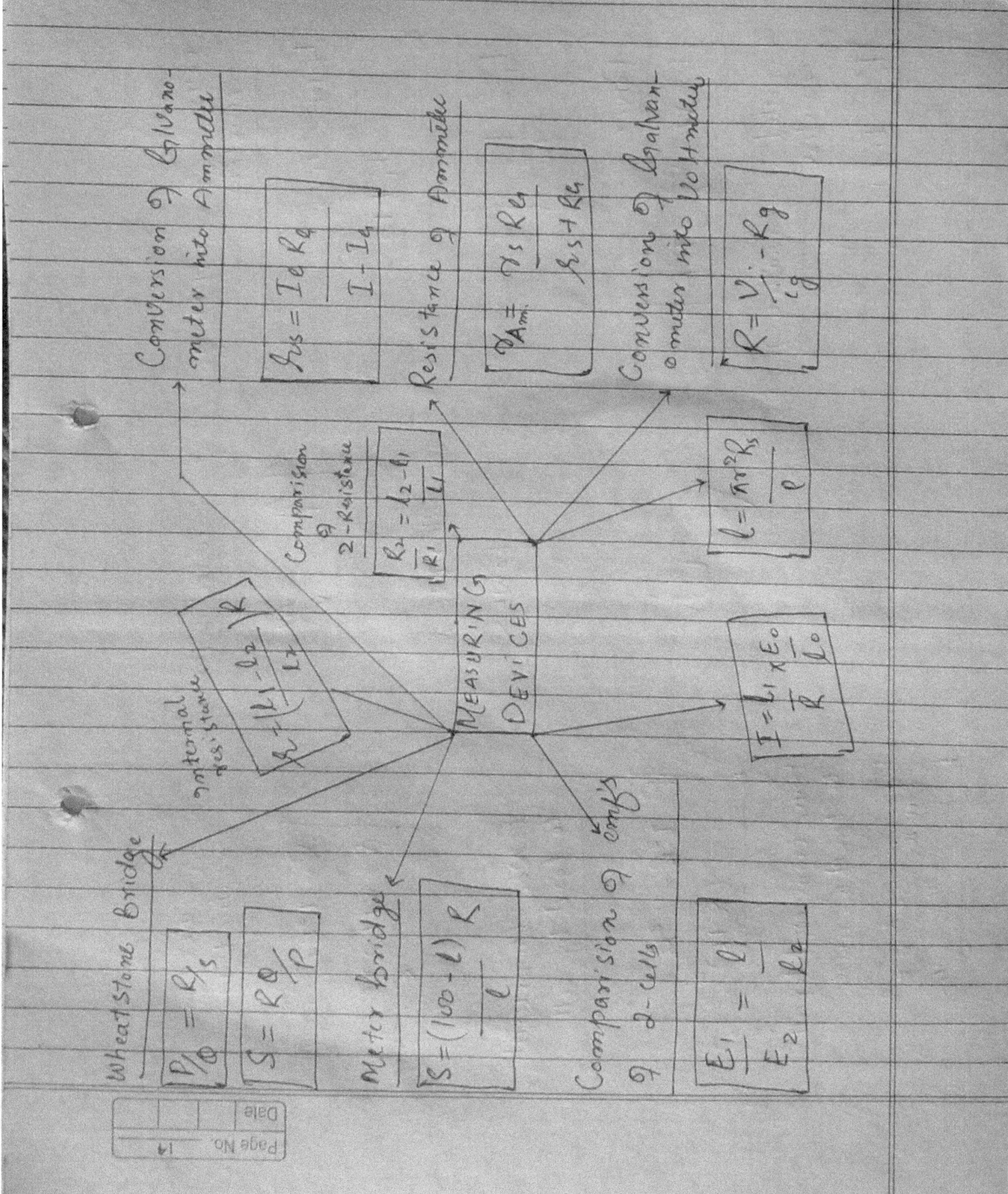

MOVING CHARGES & MAGNETISM

Biot-Savart's Law

$$dB = \frac{\mu_0}{4\pi} \frac{I\, dl \sin\theta}{r^2}$$

$$\frac{\mu_0}{4\pi} = 10^{-7} \, T\cdot m\cdot A^{-1}$$

$$\vec{dB} = \frac{\mu_0}{4\pi} \frac{I\,\vec{dl} \times \vec{r}}{r^3}$$

emf of the battery

$$v = B l /\mu_0 n$$

→ **MOVING CHARGES & MAGNETISM**

Application of Biot-Savart law's

1. Magnetic field induction due to Straight Current Carrying wire

$$B = \frac{\mu_0}{4\pi} \frac{I}{a} [\sin\phi_1 + \sin\phi_2]$$

2. Magnetic field induction due to ∞ long Straight Conductor

$$B = \frac{\mu_0}{4\pi} \frac{2I}{a} = \frac{\mu_0}{2\pi} \frac{I}{a}$$

3. Magnetic field induction at an Axial point of Circular loop

$$B = \frac{\mu_0 \, I a^2}{2(r^2+a^2)^{3/2}}$$

or

$$B = \frac{\mu_0}{4\pi} \frac{2 I a^2}{(r^2+a^2)^{3/2}}$$

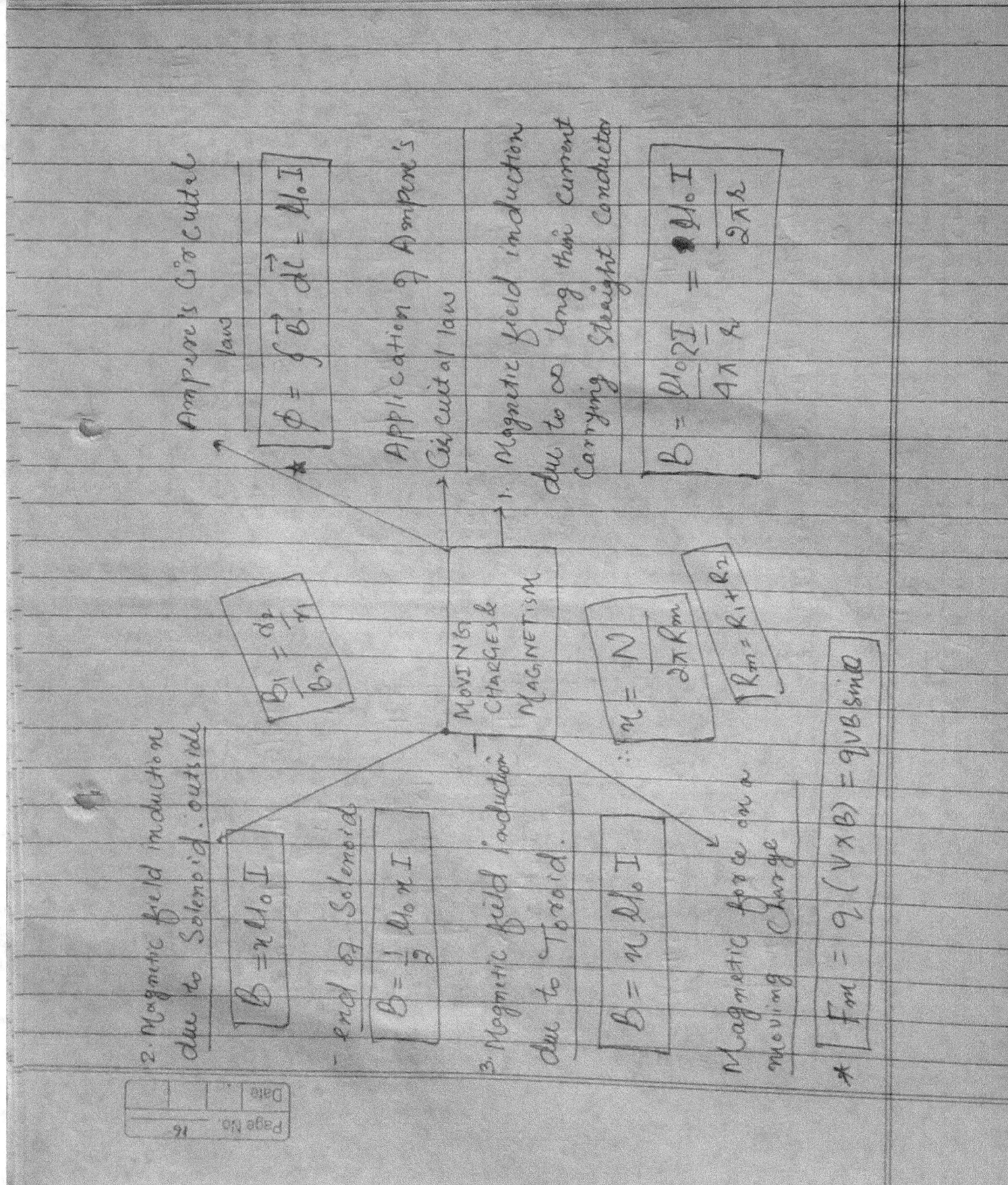

MOVING CHARGES & MAGNETISM

Ampere's Circuital law

$$\phi = \oint \vec{B} \cdot \vec{dl} = \mu_0 I$$

Application of Ampere's Circuital law

1. Magnetic field induction due to ∞ long thin current carrying straight conductor

$$B = \frac{\mu_0 2I}{4\pi R} = \frac{\mu_0 I}{2\pi R}$$

2. Magnetic field induction due to Solenoid. outside

$$B = n\mu_0 I$$

— end of Solenoid

$$B = \frac{1}{2}\mu_0 n I$$

3. Magnetic field induction due to Toroid.

$$B = n \mu_0 I$$

$$\frac{B_1}{B_2} = \frac{d_2}{d_1}$$

$$n = \frac{N}{2\pi R_m}$$

$$R_m = R_1 + R_2$$

* Magnetic force on a moving charge

$$F_m = q(\vec{v} \times \vec{B}) = qvB\sin\theta$$

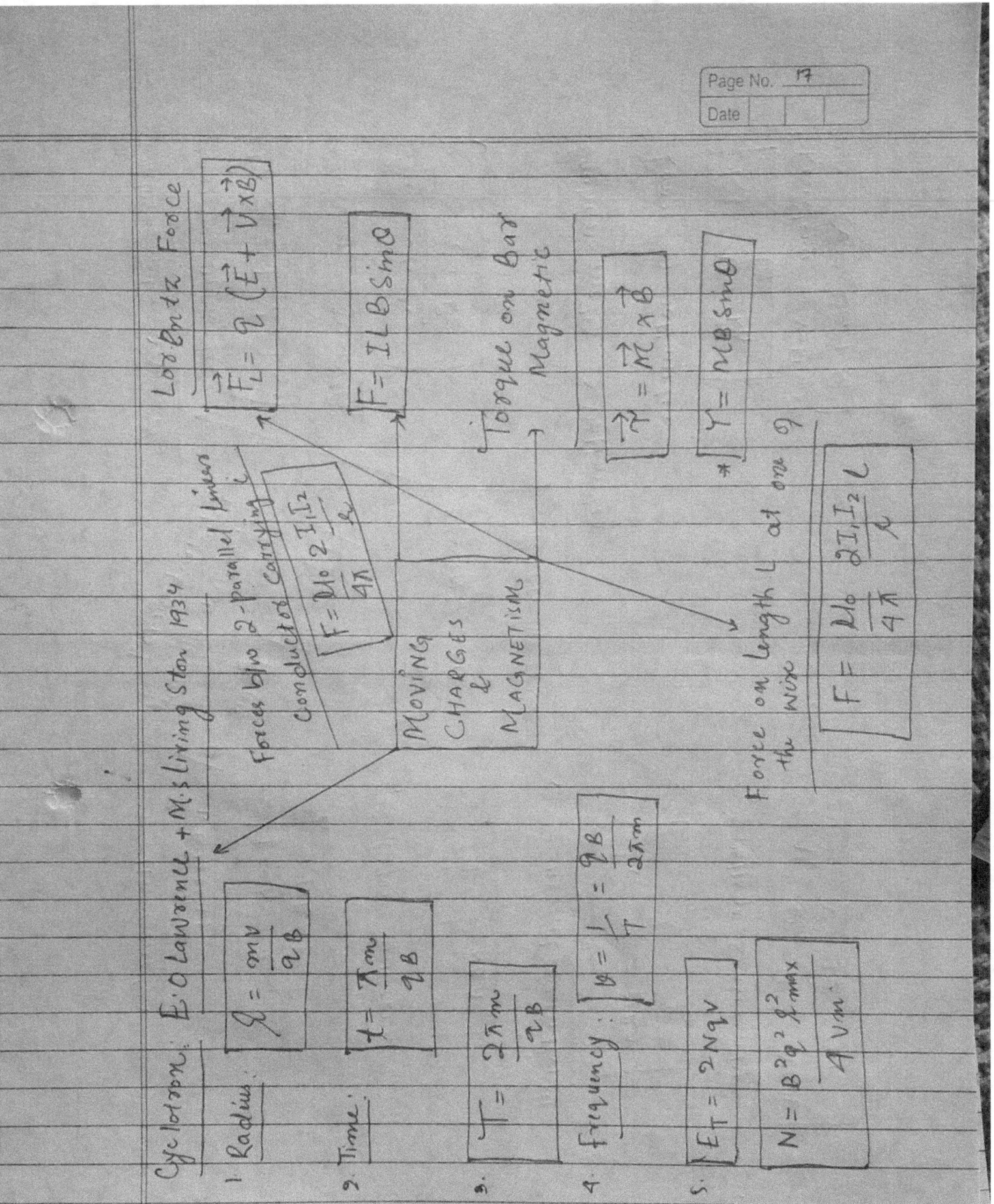

Moving Charges & Magnetism

Lorentz Force
$$\vec{F_L} = q(\vec{E} + \vec{V} \times \vec{B})$$

$$F = ILB\sin\theta$$

Torque on Bar Magnetic
$$\vec{\tau} = \vec{M} \times \vec{B}$$
$$\tau = MB\sin\theta$$

Force b/w 2-parallel wires Conductors Carrying i
$$F = \frac{\mu_0}{4\pi} \frac{2 I_1 I_2}{R}$$

Force on length L at one 9 the wire
$$F = \frac{\mu_0}{4\pi} \frac{2 I_1 I_2}{R} L$$

Cyclotron: E.O Lawrence + M.S Livingston 1934

1. **Radius:** $\quad \gamma = \frac{mV}{qB}$

2. **Time:** $\quad t = \frac{\pi m}{qB}$

3. $\quad T = \frac{2\pi m}{qB}$

4. **Frequency:** $\quad \vartheta = \frac{1}{T} = \frac{qB}{2\pi m}$

5. $E_T = 2 N q V$

$$N = \frac{B^2 q^2 R^2_{max}}{4 V m}$$

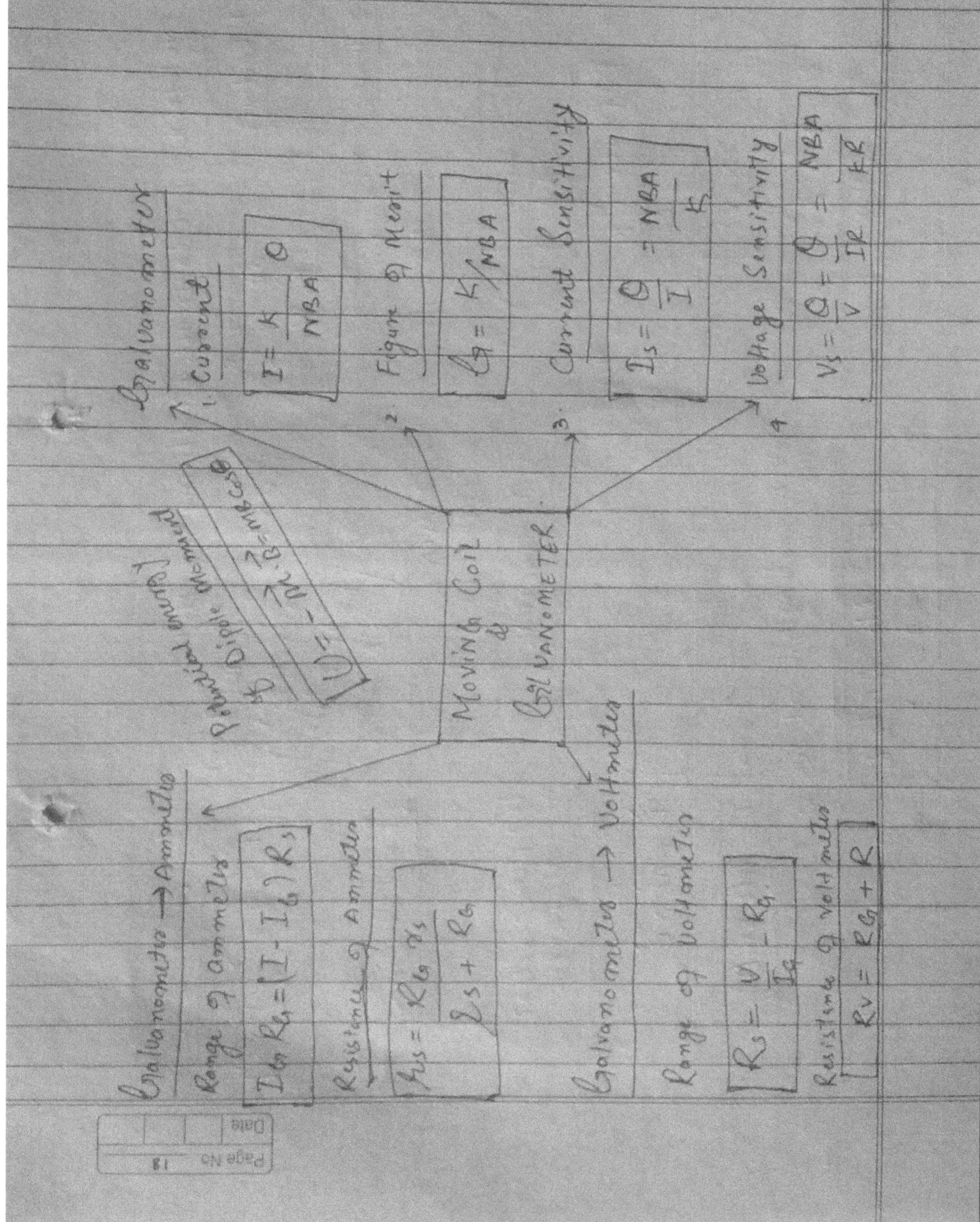

MAGNETISM AND MATTER

MAGNETISM AND MATTER

Geo-Magnetism
Dr. William Gilbert

Horizontal Component of Earth magnetic field:
$$B_H = B \cos\theta$$

Vertical:
$$B_V = B \sin\theta$$

$$\tan\theta = \frac{B_V}{B_H}$$

$$B = \sqrt{B_H^2 + B_V^2}$$

Comparison of B at 2-different places
$$\frac{B_1}{B_2} = \frac{T_2^2 \cos\theta}{T_1^2 \cos\theta}$$

Comparison of Magnetic Moments of the Same size
$$\frac{M_1}{M_2} = \frac{T_2^2}{T_1^2}$$

Comparison of Magnetic Moments of different size
$$\frac{T_1}{T_2} = \sqrt{\frac{M_1 - M_2}{M_1 + M_2}}$$

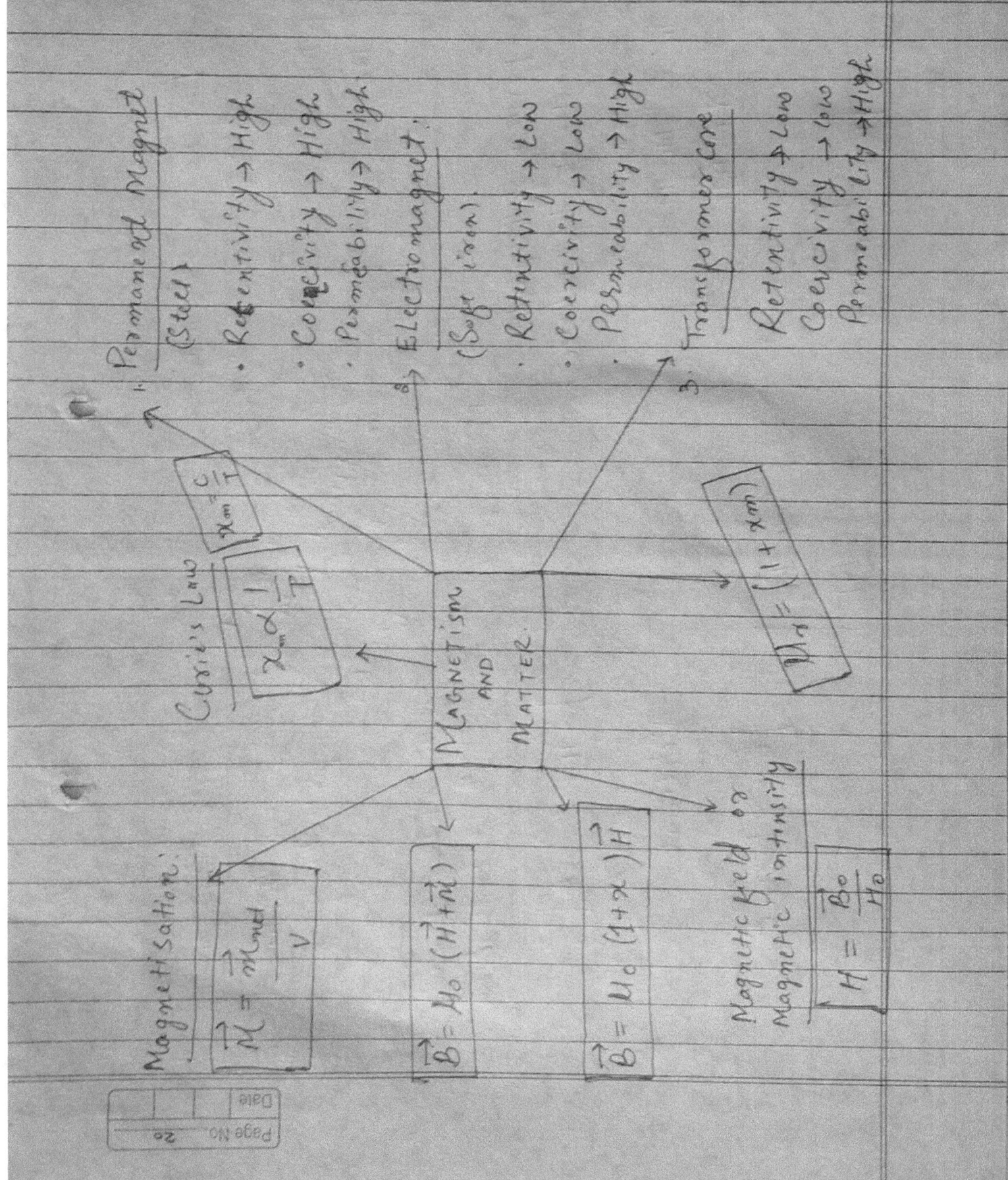

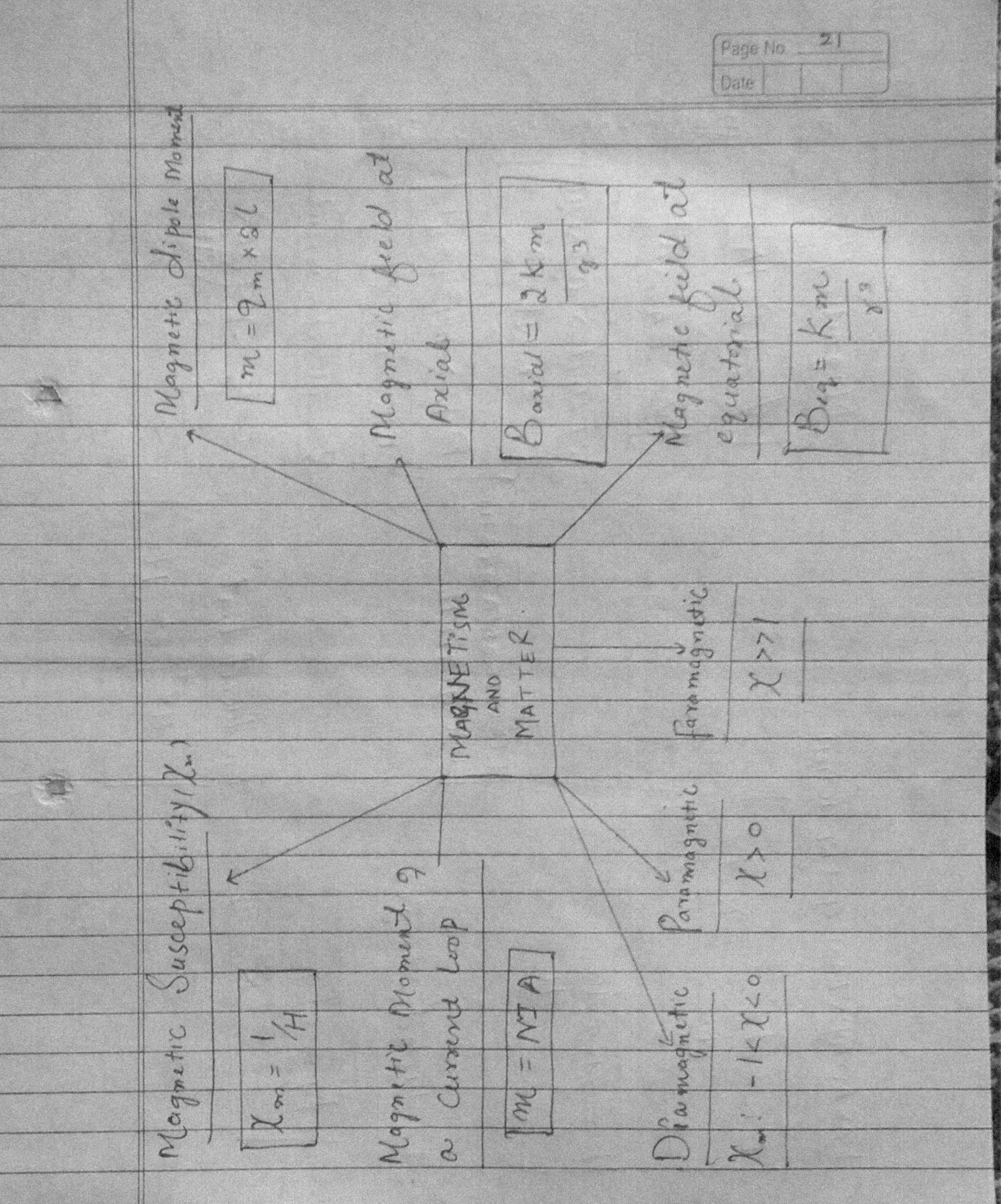

ELECTROMAGNETIC INDUCTION (EMI)

Magnetic flux

$$\phi = \oint \vec{B} \cdot d\vec{A}$$

$$\phi = BA\cos\theta$$

SI unit → Maxwell or $Wb\,m^{-2}$

D.f → $M^1 L^2 T^{-2} A^{-1}$

$1\,Wb = 10^8\,Mx$

Induced Current

$$I = \frac{E}{R} = \frac{-1}{R}\left(\frac{d\phi}{dt}\right)$$

Electromagnetic Induction

Induced Heat

$$H = \int_0^t \frac{e^2}{R} dt$$

Faraday's Law of Induction

$$E \propto -\frac{d\phi}{dt}$$

$$E = -N\frac{d\phi}{dt}$$

Induced Charge

$$dq = \frac{d\phi}{R}$$

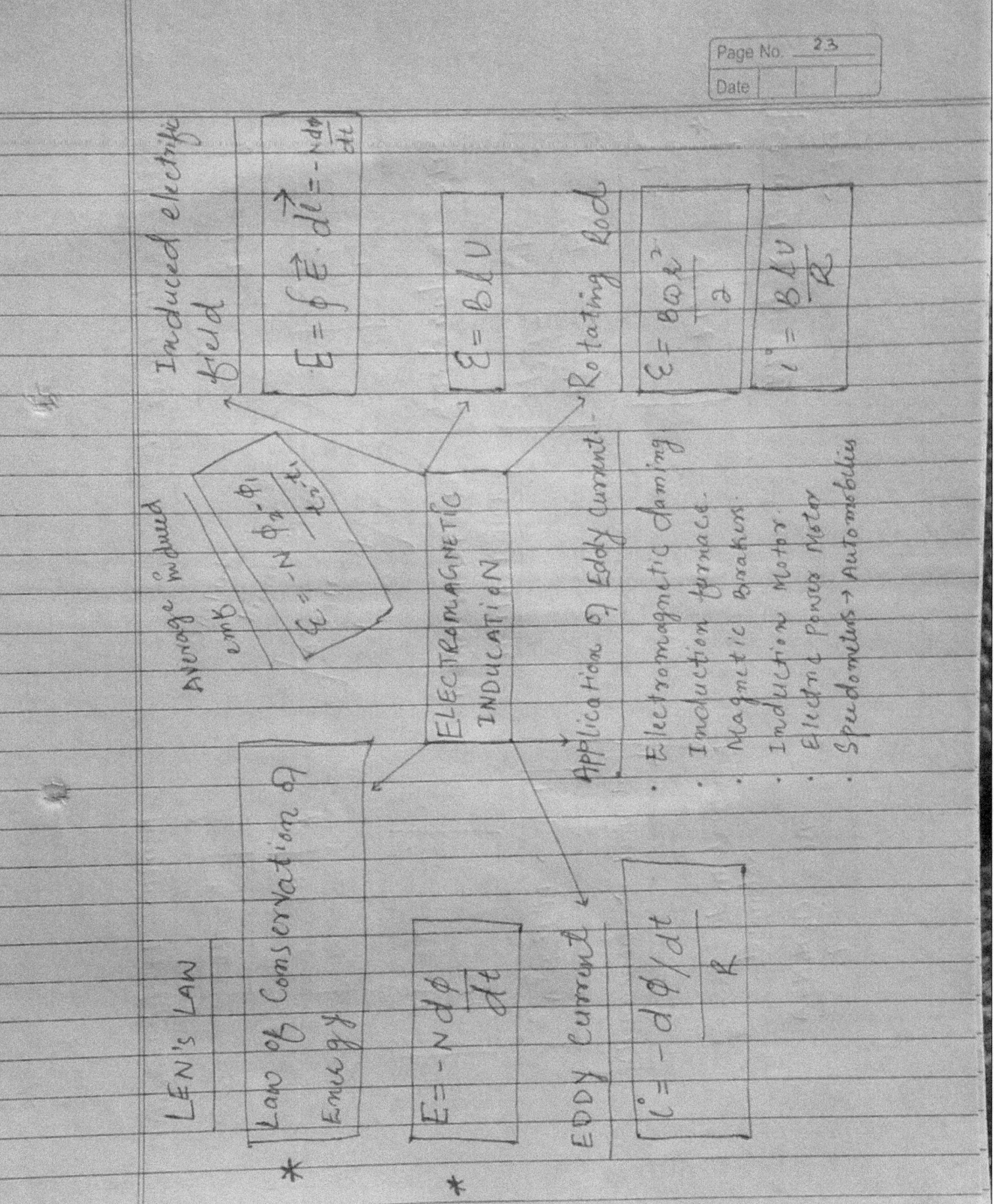

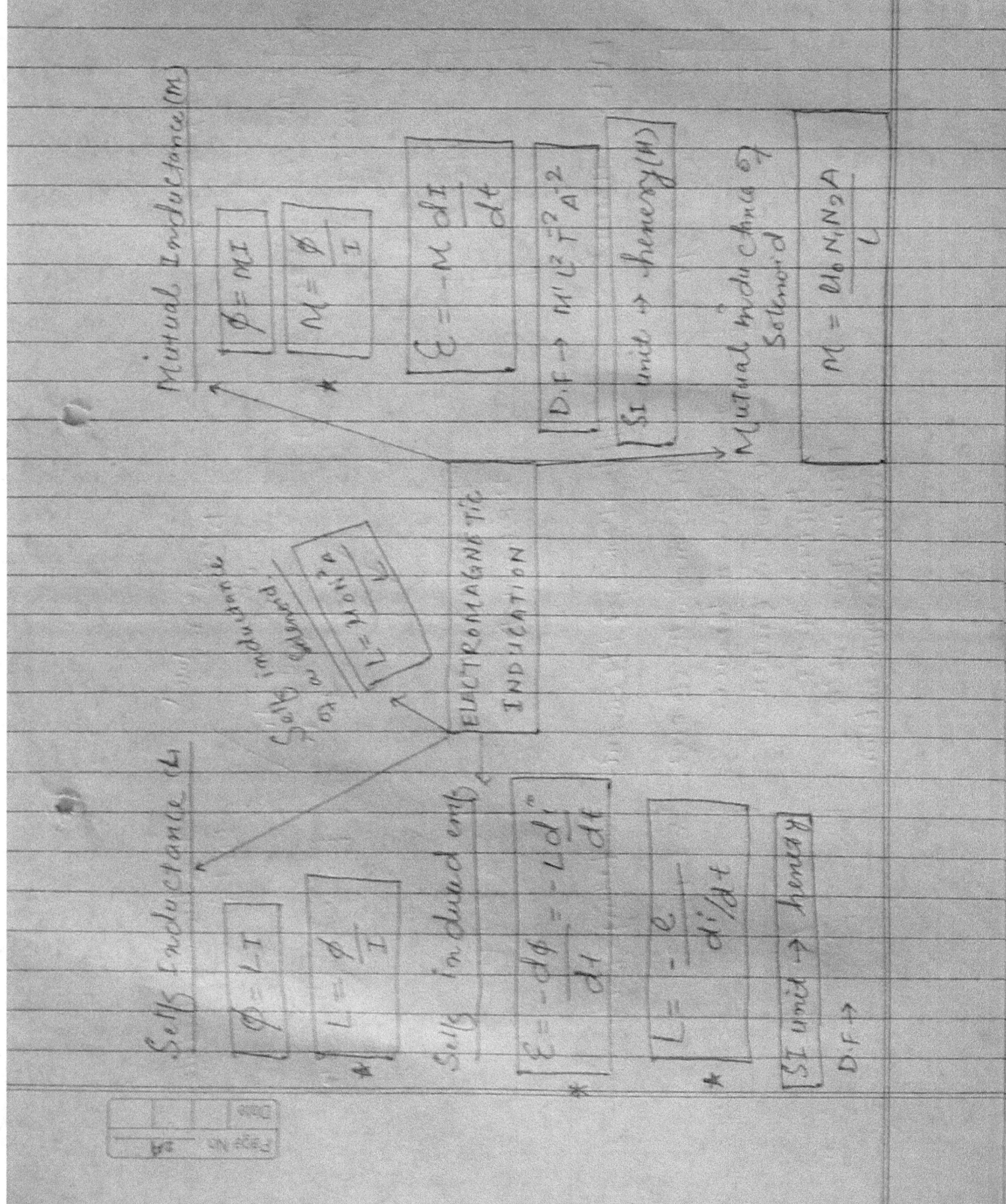

ELECTROMAGNETIC INDUCTION

Self Inductance (L)

$[\phi = LI]$

$[L = \dfrac{\phi}{I}]$

Self induced emf

$*\ \left[\varepsilon = -\dfrac{d\phi}{dt} = -L\dfrac{di}{dt}\right]$

$*\ \left[L = -\dfrac{\varepsilon}{di/dt}\right]$

SI unit → henry

D.F →

Self inductance of a solenoid

$\left[L = \dfrac{\mu_0 N^2 A}{l}\right]$

Mutual Inductance (M)

$[\phi = MI]$

$[M = \dfrac{\phi}{I}]$

$*\ \left[\varepsilon = -M\dfrac{dI}{dt}\right]$

$[D.F \to M^1 L^2 T^{-2} A^{-2}]$

$[SI\ unit \to henry(H)]$

Mutual inductance of solenoid

$M = \dfrac{\mu_0 N_1 N_2 A}{l}$

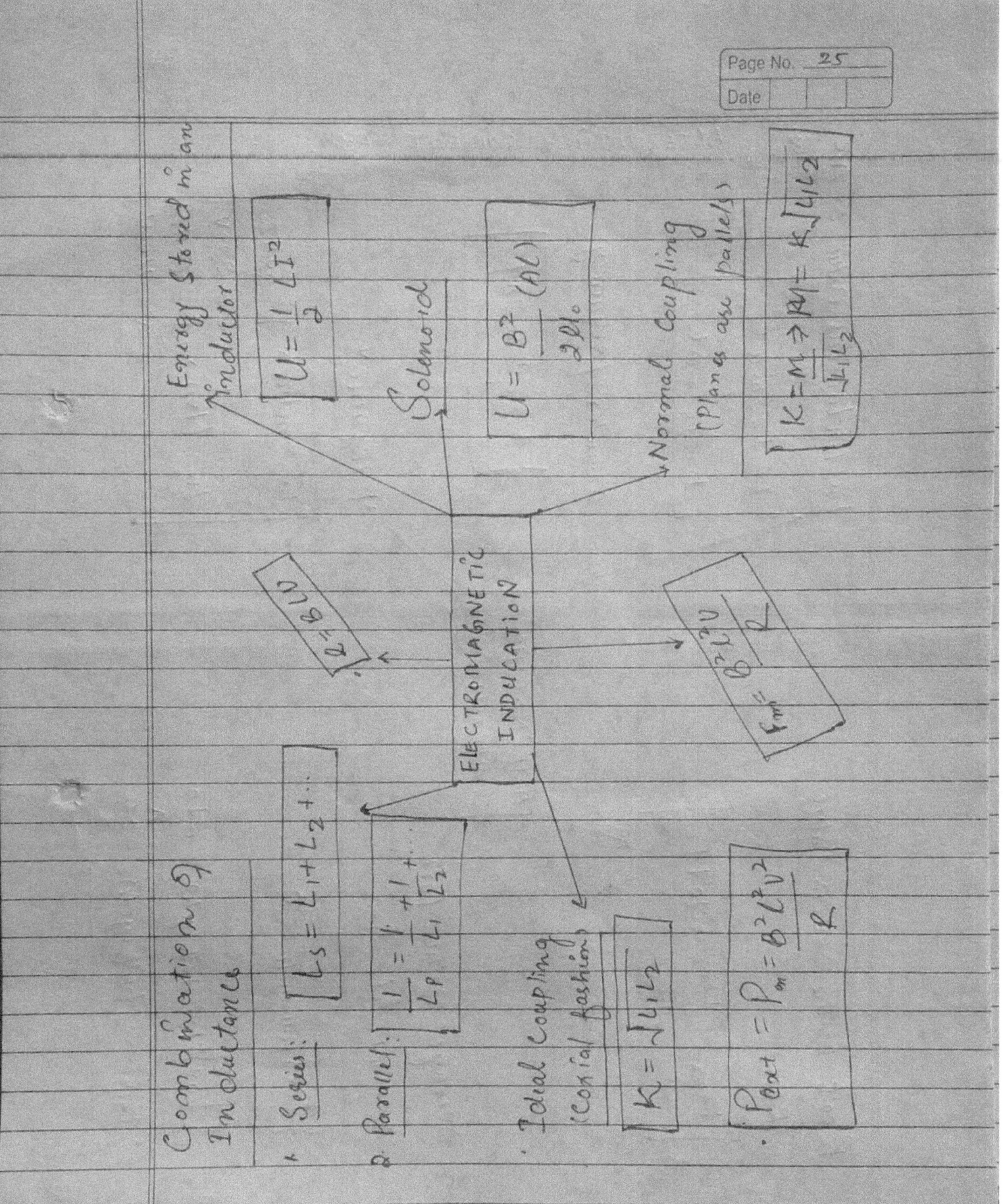

ALTERNATING CURRENT (A.C).

ALTERNATING CURRENT

* Instantaneous value of A.C

$$I = I_0 \sin \omega t$$
or
$$I = I_0 \cos \omega t$$

Angular frequency

$$\omega = \frac{2\pi}{T} = 2\pi f$$

Frequency

In India → $f = 50 Hz$, $v = 220 V$
In USA → $f = 60 Hz$, $v = 110 V$

* Root Mean Square Value (RMS)

$$I_{rms} = \frac{I_0}{\sqrt{2}} = 0.707 I_0$$

Average Value for Sinusoidal A.C

$$\{I_{av}\}_{Full\ cycle} = 0$$

$$\{I_{av}\}_{+ve\ half\ cycle} = \frac{2I_0}{\pi}$$

$$\{I_{av}\}_{-ve\ half\ cycle} = -\frac{2I_0}{\pi}$$

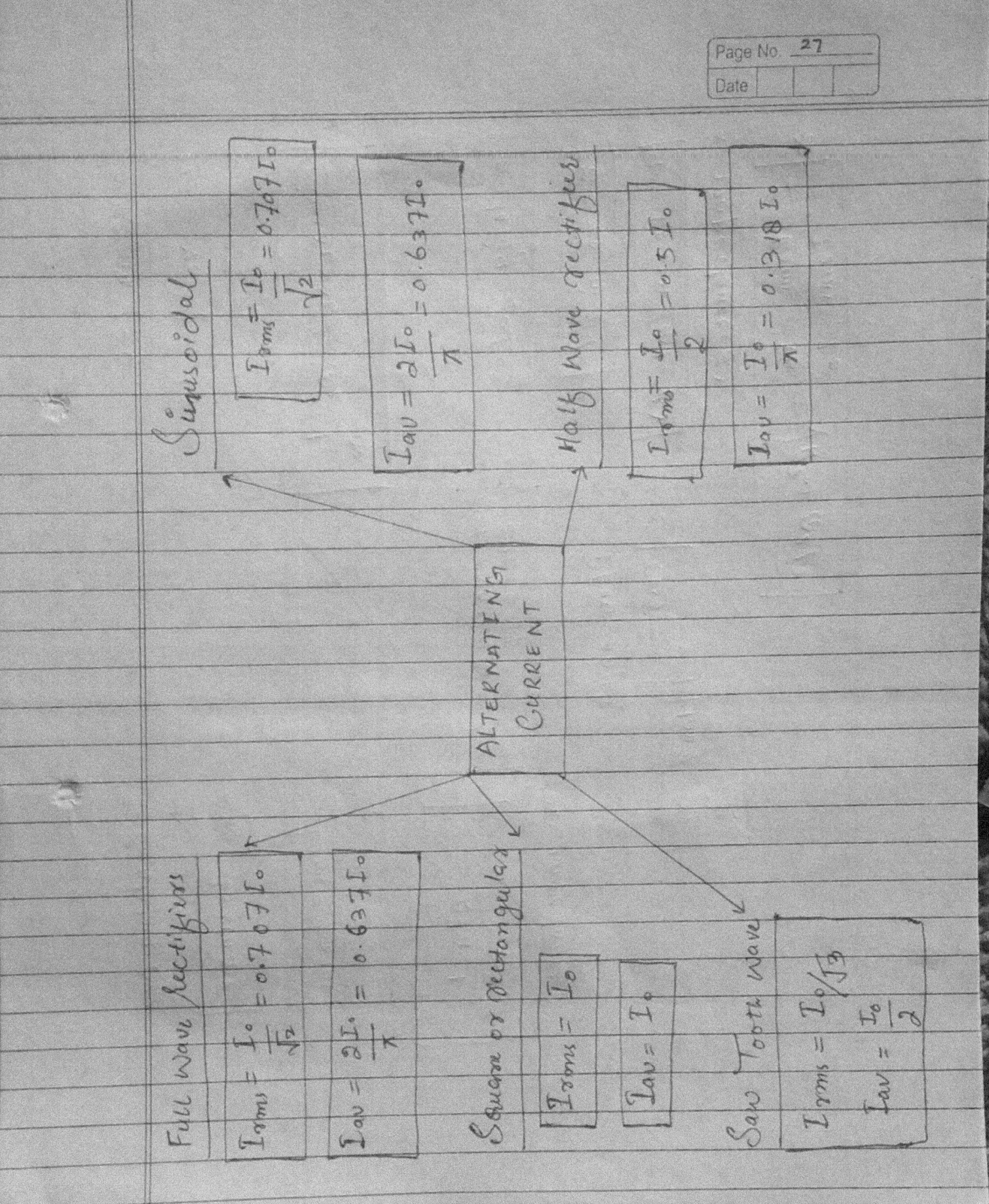

ALTERNATING CURRENT

Sinusoidal
$$I_{rms} = \frac{I_0}{\sqrt{2}} = 0.707 I_0$$
$$I_{av} = \frac{2I_0}{\pi} = 0.637 I_0$$

Half Wave rectifier
$$I_{rms} = \frac{I_0}{2} = 0.5 I_0$$
$$I_{av} = \frac{I_0}{\pi} = 0.318 I_0$$

Full Wave Rectifier
$$I_{rms} = \frac{I_0}{\sqrt{2}} = 0.707 I_0$$
$$I_{av} = \frac{2I_0}{\pi} = 0.637 I_0$$

Square or Rectangular
$$I_{rms} = I_0$$
$$I_{av} = I_0$$

Saw Tooth Wave
$$I_{rms} = I_0/\sqrt{3}$$
$$I_{av} = \frac{I_0}{2}$$

ALTERNATING CURRENT

Containing R-C

$$I = I_0 \sin \omega t$$

$$I_{rms} = \frac{E_{o\,rms}}{Z}$$

$$Z = \sqrt{R^2 + \frac{1}{\omega^2 C^2}}$$

* Z = impedance

$$\tan \phi = \frac{X_C}{R}$$

- Phase angle

$$\cos \phi = \frac{R}{Z}$$

Containing L-C

$$I_0 = \frac{V_0}{Z}$$

$$Z = X_L - X_C = \omega L - \frac{1}{\omega C}$$

Containing R-L

$$\varepsilon = \varepsilon_0 \sin \omega t$$

$$I = I_0 \sin(\omega t - \phi)$$

* impedance (Z)

$$Z = \frac{E_{rms}}{I_{rms}} = \frac{V_0}{I_0} = \sqrt{R^2 + \omega^2 L^2}$$

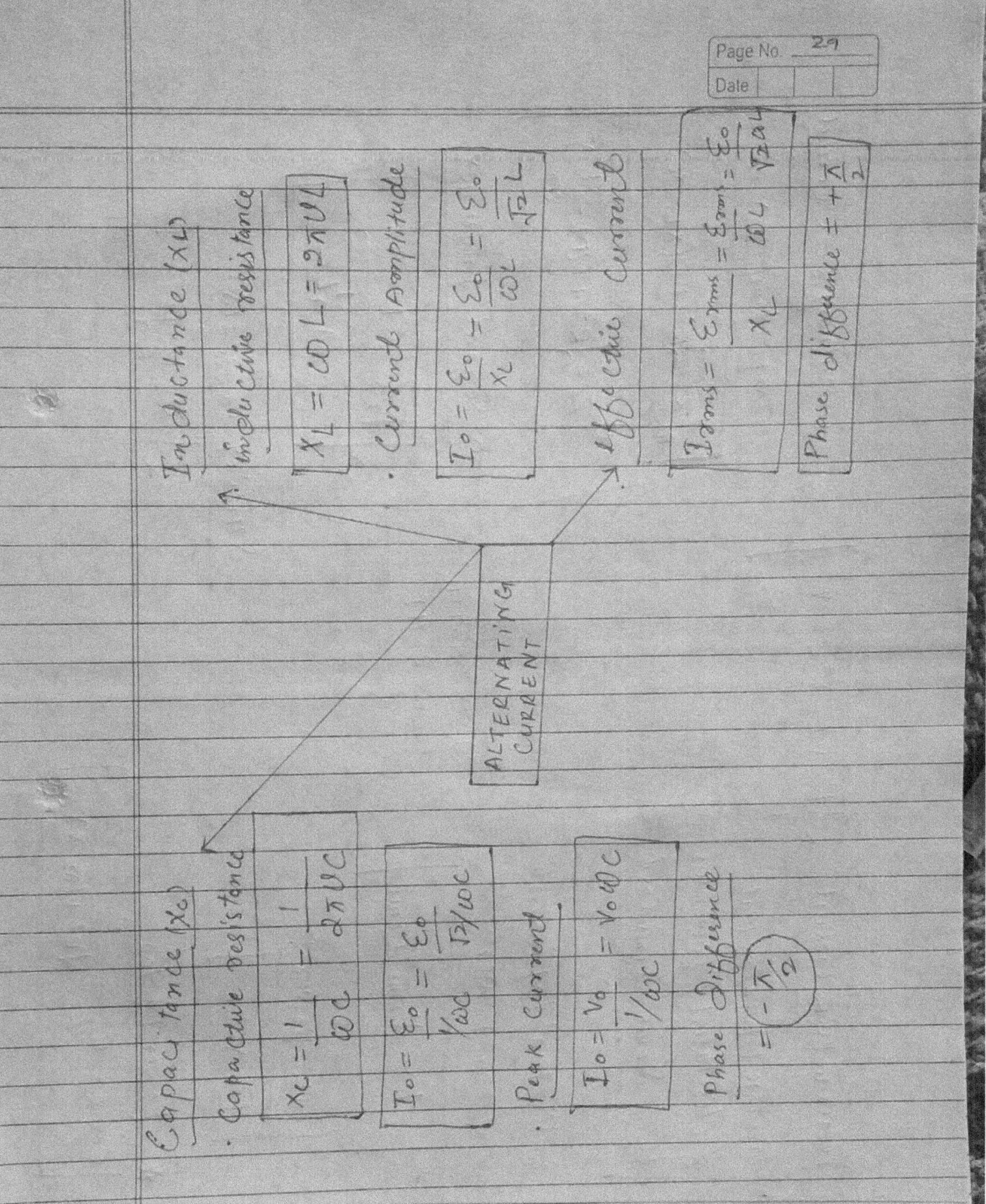

ALTERNATING CURRENT

Inductance (X_L)

- Inductive resistance

$$X_L = \omega L = 2\pi \nu L$$

- Current Amplitude

$$I_0 = \frac{\varepsilon_0}{\omega L} = \frac{\varepsilon_0}{X_L} = \frac{\varepsilon_0}{2\pi \nu L}$$

- Effective Current

$$I_{rms} = \frac{\varepsilon_{rms}}{X_L} = \frac{\varepsilon_0}{\omega L \sqrt{2}} = \frac{\varepsilon_0}{\sqrt{2} \omega L}$$

Phase difference $= +\dfrac{\pi}{2}$

Capacitance (X_C)

- Capacitive resistance

$$X_C = \frac{1}{\omega C} = \frac{1}{2\pi \nu C}$$

$$I_0 = \frac{\varepsilon_0}{1/\omega C} = \frac{\varepsilon_0}{1/\omega C}$$

- Peak Current

$$I_0 = \frac{V_0}{1/\omega C} = V_0 \omega C$$

Phase difference
$= -\pi/2$

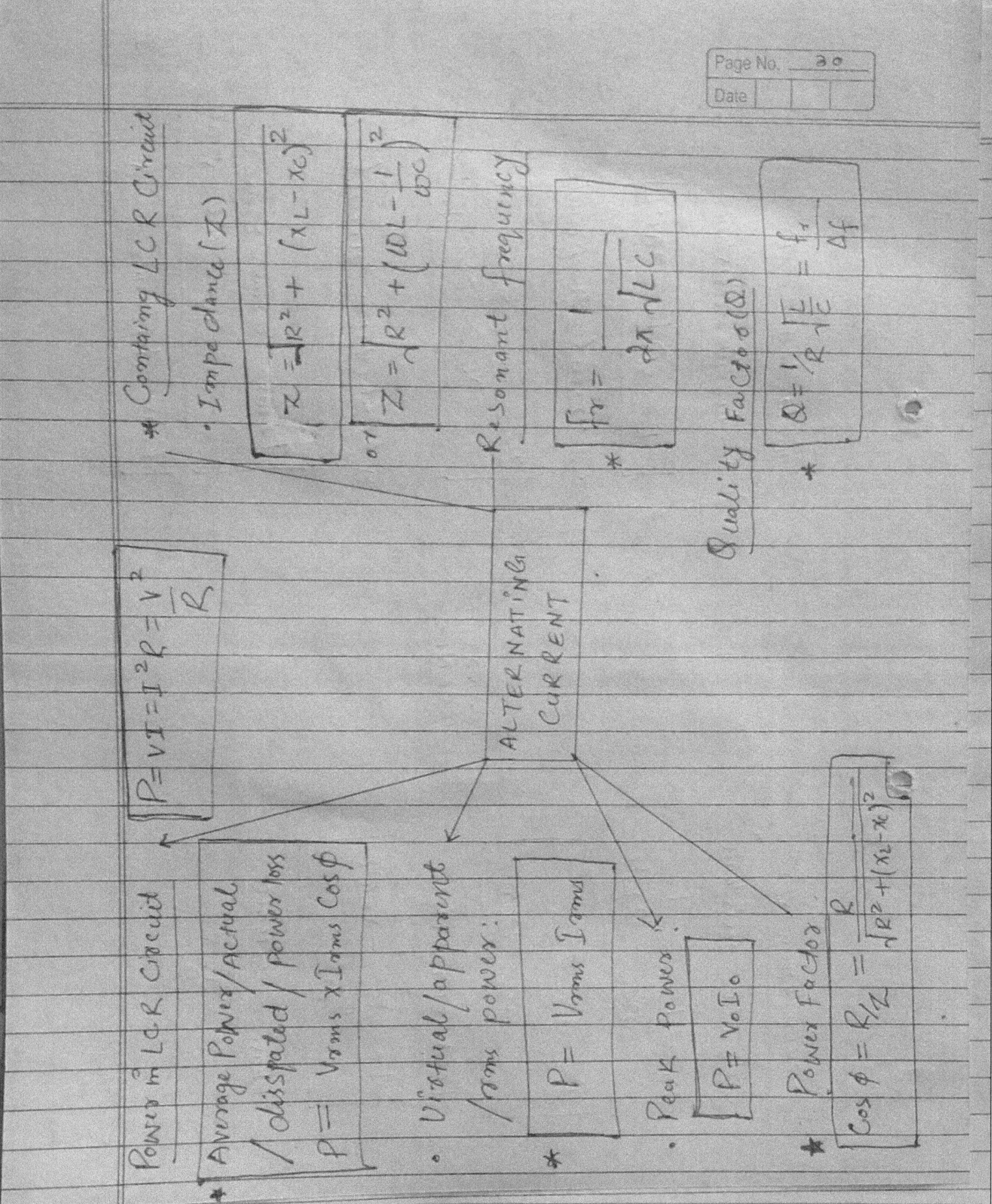

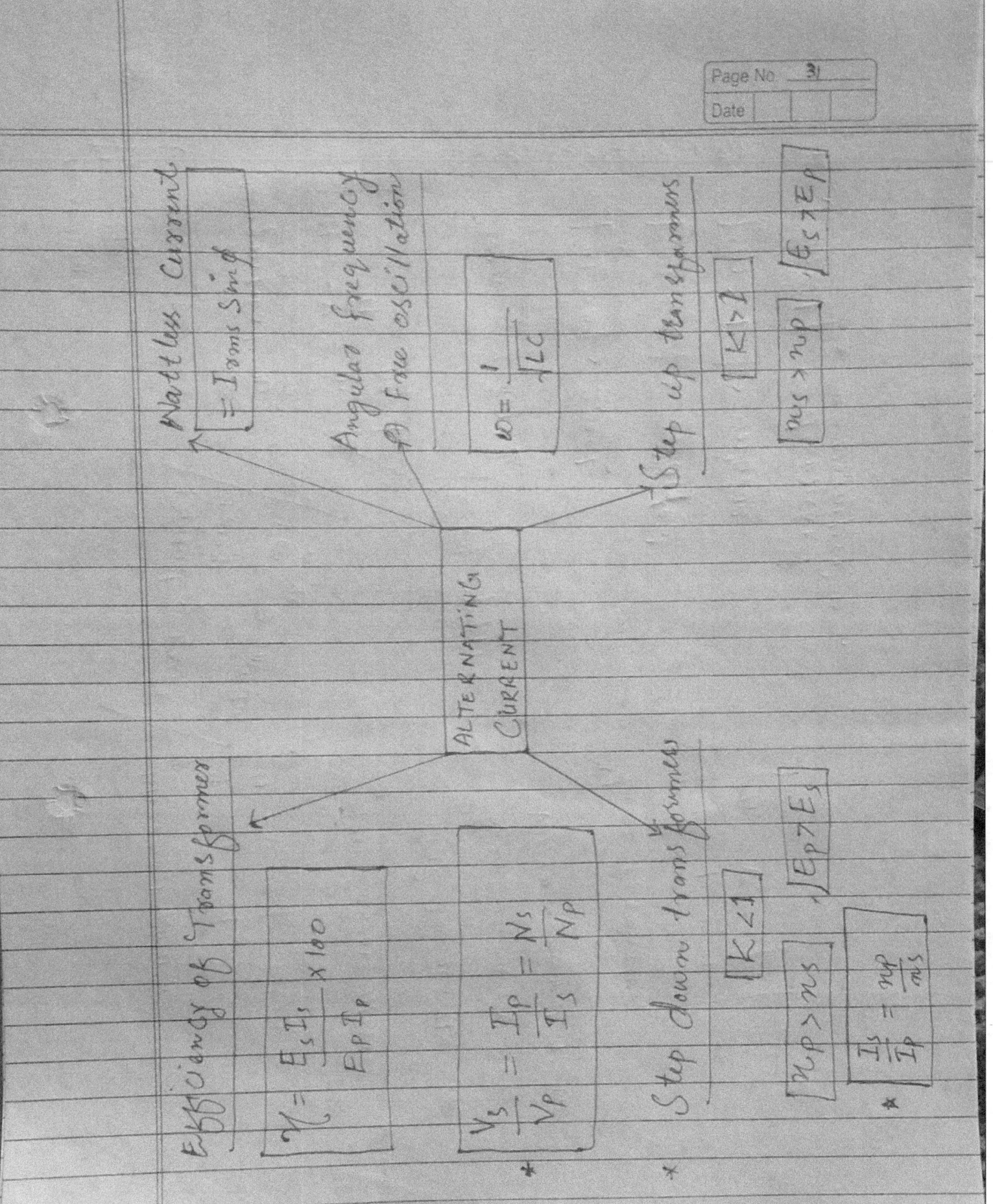

★ ELECTROMAGNETIC WAVES:

Displacement Current

$$I_D = \varepsilon_0 \frac{d\phi_E}{dt}$$

$$I_D = \varepsilon_0 A \frac{dE}{dt}$$

$$I_D = \varepsilon_0 A \frac{d}{dt}\left(\frac{V}{d}\right) = \frac{\varepsilon_0 A}{d}\frac{dV}{dt}$$

$$I_D = C\frac{dV}{dt}$$

Modified Ampere's Circuital law:

$$\oint \vec{B} \cdot \vec{dl} = \mu_0 (I + I_D)$$

$$\oint \vec{B} \cdot \vec{dl} = \mu_0 \left(I + \varepsilon_0 \frac{d\phi_E}{dt}\right)$$

ELECTROMAGNETIC WAVES

$$\frac{E_0}{B_0} = C$$

$$B_0 = \frac{E_0}{C}$$

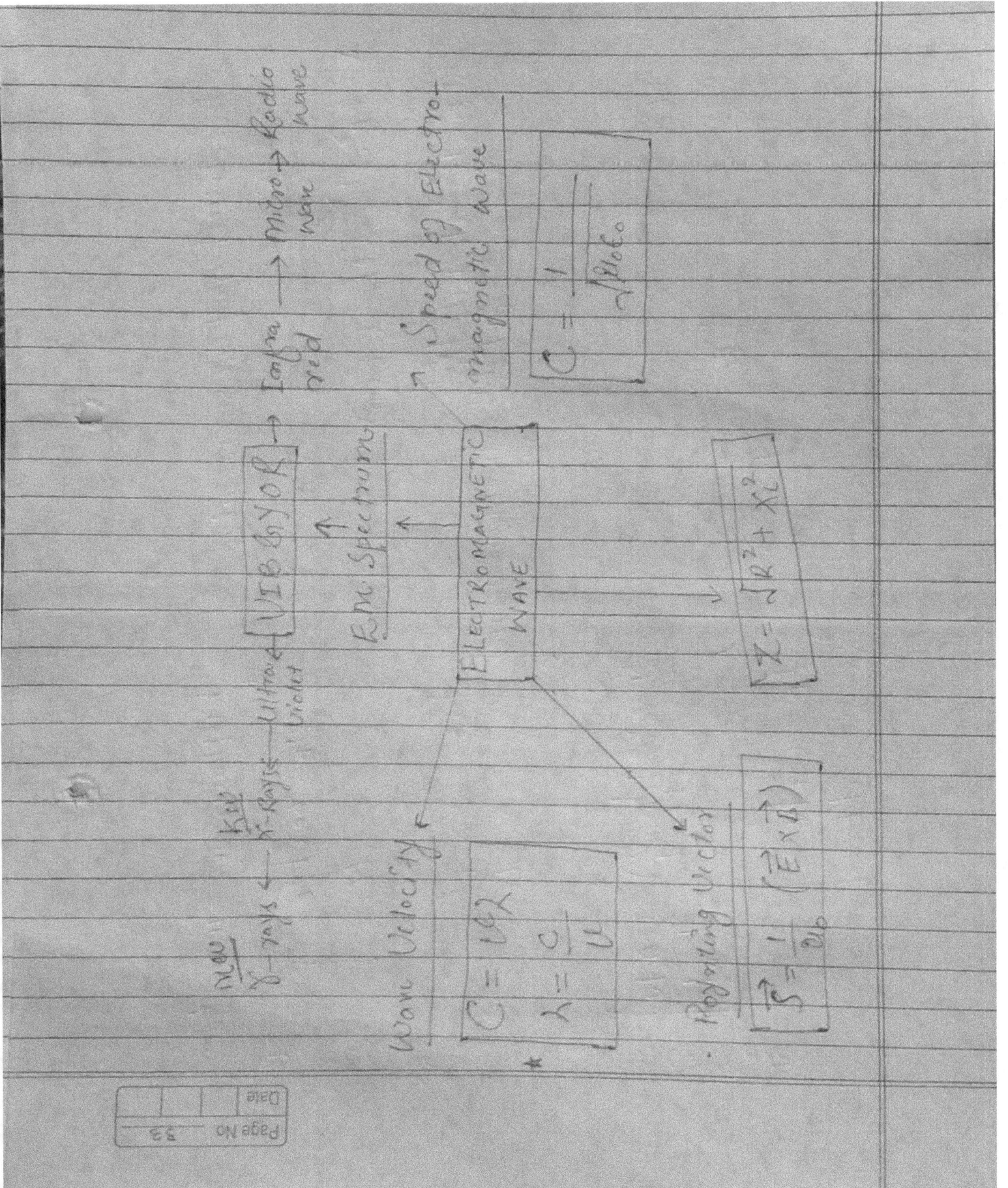

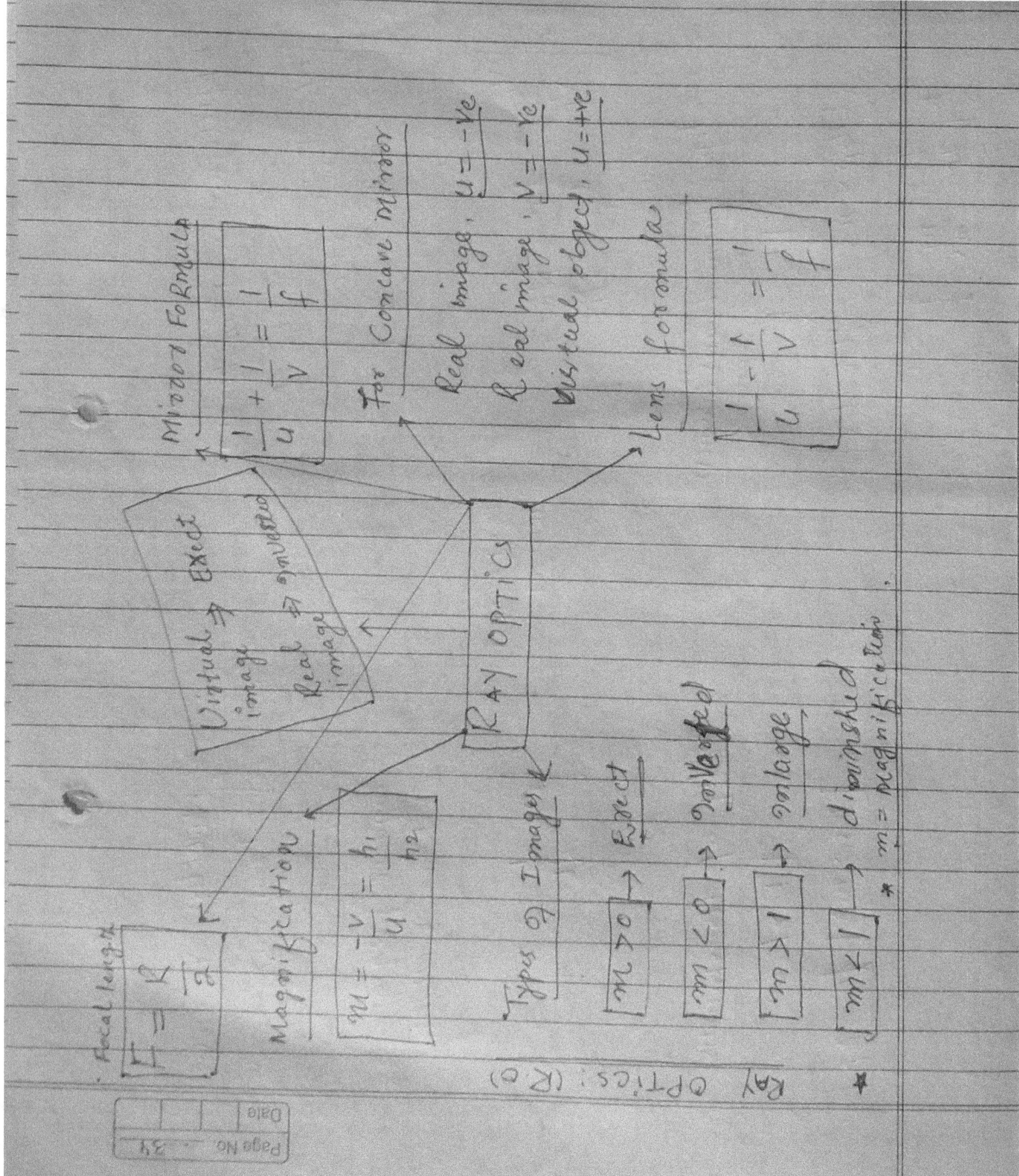

Ray Optics (R.O)

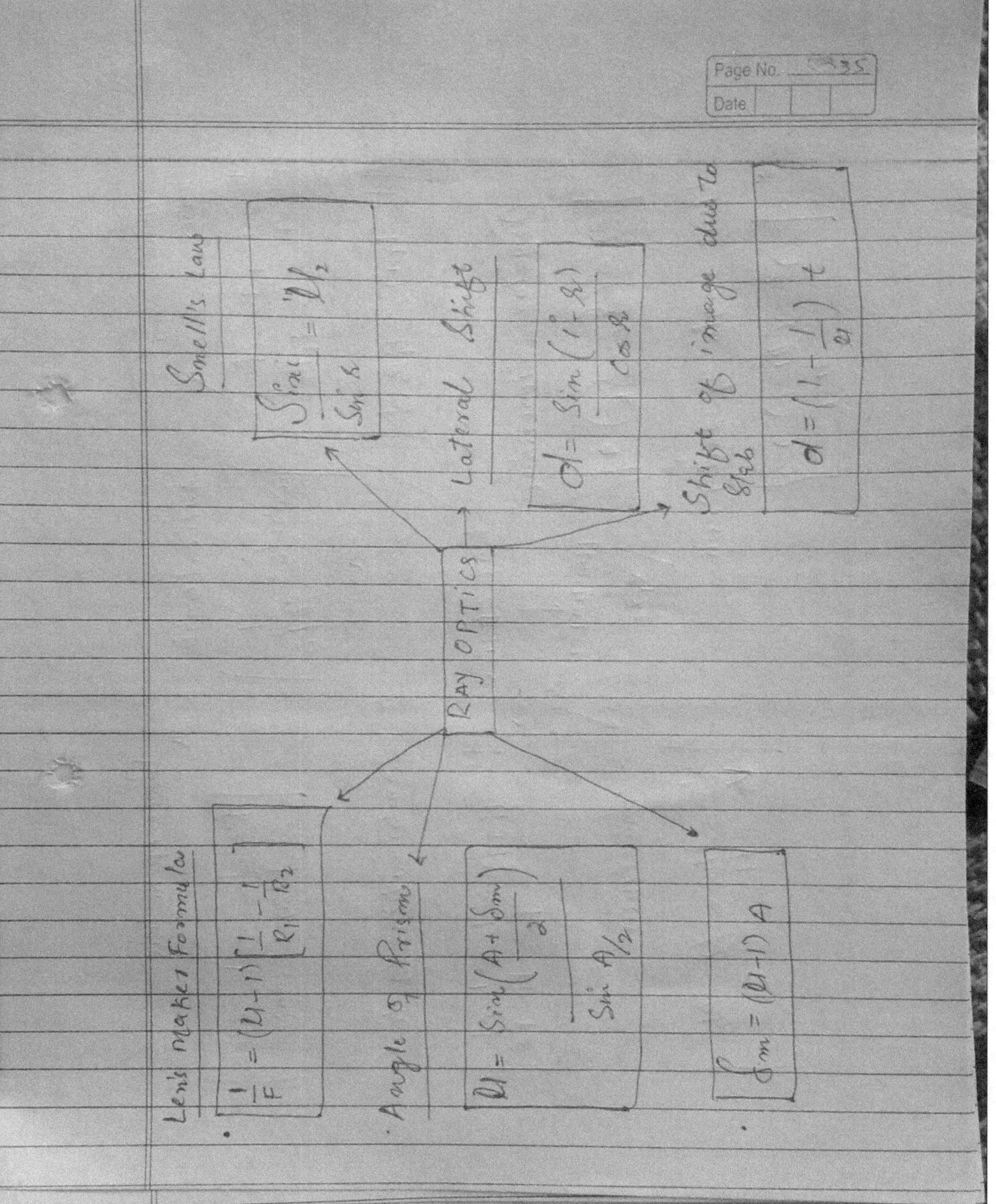

RAY OPTICS

Snell's Law

$$\frac{\sin i}{\sin r} = \frac{n_2}{n_1}$$

Lateral Shift

$$d = \frac{\sin(i - r)}{\cos r}$$

Shift of image due to Slab

$$d = \left(1 - \frac{1}{n}\right) t$$

Lens maker Formula

$$\frac{1}{F} = (n-1)\left[\frac{1}{R_1} - \frac{1}{R_2}\right]$$

Angle of Prism

$$n = \frac{\sin\left(\frac{A + \delta_m}{2}\right)}{\sin A/2}$$

$$\delta_m = (n-1) A$$

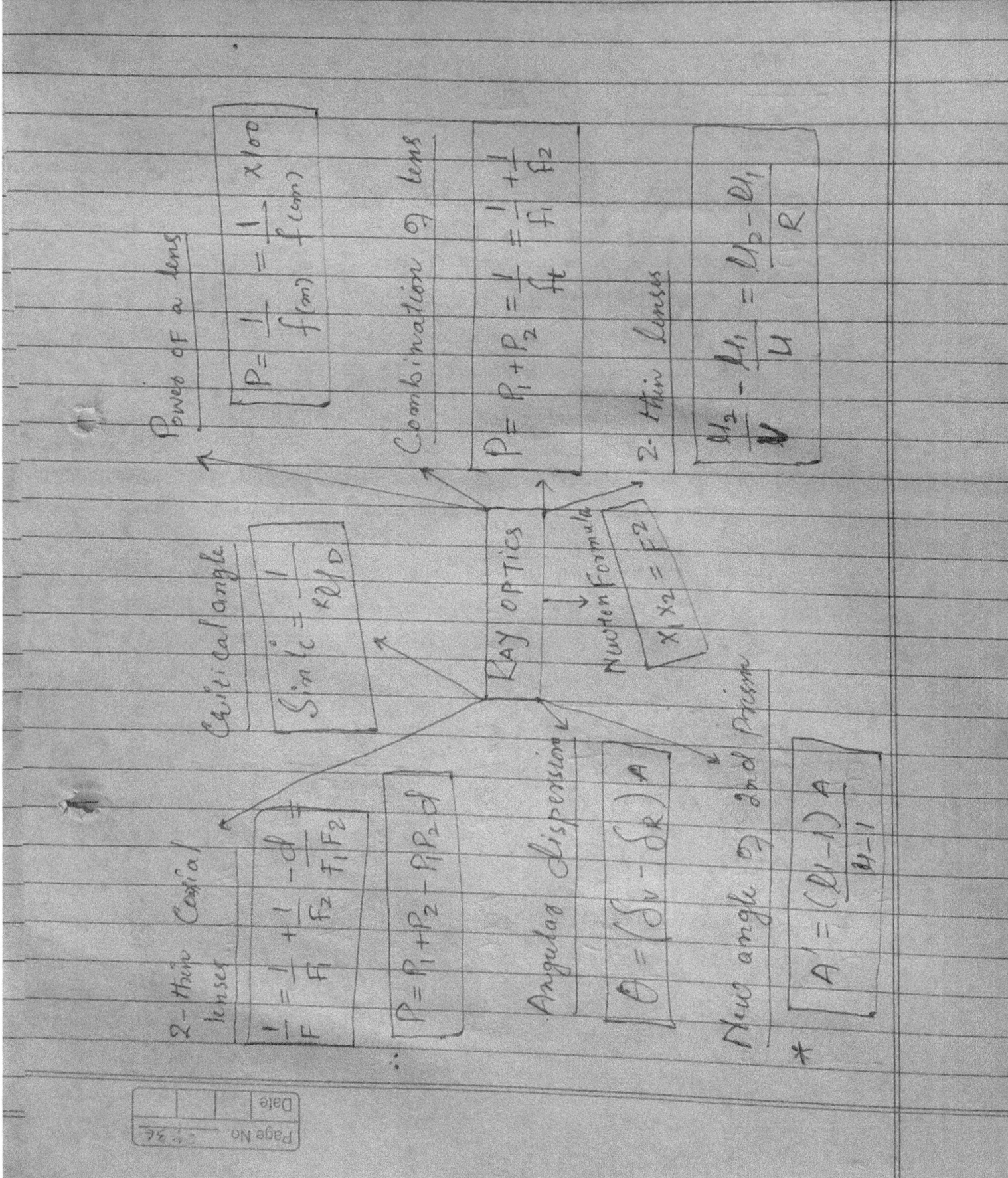

RAY OPTICS

2-thin Coaxial lenses
$$\frac{1}{F} = \frac{1}{F_1} + \frac{1}{F_2} - \frac{d}{F_1 F_2}$$
$$P = P_1 + P_2 - P_1 P_2 d$$

Critical angle
$$\sin i_c = \frac{1}{R.I.D}$$

Power of a lens
$$P = \frac{1}{f(m)} = \frac{1}{f(cm)} \times 100$$

Combination of lens
$$P = P_1 + P_2 = \frac{1}{ft} = \frac{1}{f_1} + \frac{1}{f_2}$$

2-thin lenses
$$\frac{\mu_2}{V} - \frac{\mu_1}{U} = \frac{\mu_2 - \mu_1}{R}$$

Newton Formula
$$x_1 x_2 = F^2$$

Angular dispersion
$$\theta = (\delta_V - \delta_R) A$$

New angle of 2nd Prism
$$A' = \frac{(\mu-1) A}{\mu'-1}$$

OPTICAL INSTRUMENTS

Simple Microscope

$$M_P = \beta/\alpha$$

- image formed at least distance

$$M = 1 + D/f$$

- image formed at ∞

$$M = D/f$$

Compound Microscope

$$M_P = m_o m_e$$

- Image at least distance

$$M = \frac{v_o}{u_o}\left(1 + \frac{D}{f_e}\right)$$

- Image at ∞

$$M = \frac{v_o}{u_o}\left(\frac{D}{f_e}\right)$$

$$L = v_o + f_e$$

$$L = v_o + \frac{f_e D}{f_e + D}$$

OPTICAL INSTRUMENTS

Astronomical (Refracting type)

- Image at least distance

$$M = \frac{f_o}{f_e}\left(1 + \frac{f_e}{D}\right)$$

- Image at ∞

$$M = \frac{f_o}{f_e}$$

$$L = f_o + f_e$$

WAVE OPTICS

Source of Light → Point
Source - Wavefront → Spherical

$$A \propto \frac{1}{x}$$

$$I \propto \frac{1}{x^2} \propto A^2$$

Source of Light → Linear
Wavefront → Cylindrical

$$A \propto \frac{1}{\sqrt{x}}$$

$$I \propto A^2 \propto \frac{1}{x}$$

Constructive interference

Path difference:

$$x_n = \frac{n\lambda D}{d} = n\lambda$$

Phase difference:

$$\phi = 2n\pi$$

$$I_{max} = (\sqrt{I_1} + \sqrt{I_2})^2 = (A_1 + A_2)^2$$

$$A_{max} = A_1 + A_2$$

$$I_{max} = 4I_0$$

Wave Optics

Destructive interference

$$I^o_{min} = 0$$

- Path difference
$$x = (2m-1)\frac{\lambda}{2}$$

- Phase difference
$$\phi = (2m-1)\pi$$

$$I_{min} = (\sqrt{I_1} - \sqrt{I_2})^2 = (A_1 - A_2)^2$$

$$A_{min} = A_1 - A_2$$

$$I_{max} = (\sqrt{I_1} + \sqrt{I_2})^2 = (A_1 + A_2)^2$$

Fringe width

$$\beta = \frac{\lambda D}{d}$$

$$\beta' = \frac{\lambda' \beta}{\lambda}$$

Angular width

$$\theta = \frac{\beta}{D} = \frac{\lambda}{d}$$

$$\theta' = \theta/\mu$$

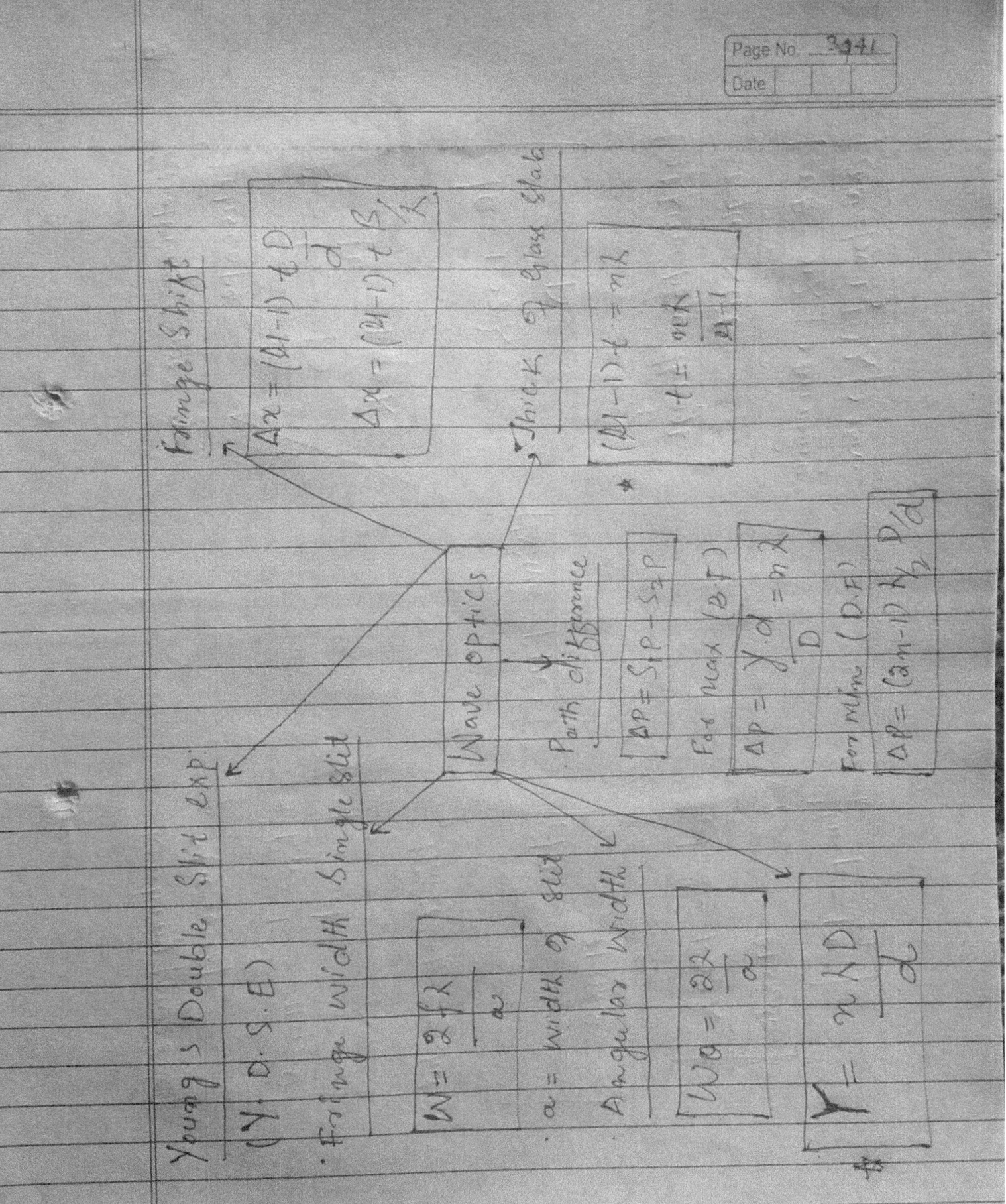

Wave Optics

Young's Double Slit Exp. (Y.D.S.E)

- Fringe Width

$$W = \frac{\lambda D}{d}$$

a = width of slit

- Angular Width

$$W_a = \frac{2\lambda}{d}$$

$$Y = \frac{n \lambda D}{d}$$

Single Slit

Path difference

$$\Delta P = S_1 P - S_2 P$$

For Max (B.F)
$$\Delta P = \frac{y \cdot d}{D} = n\lambda$$

For Min (D.F)
$$\Delta P = (2n-1) \frac{\lambda}{2} \cdot D/d$$

Fringe Shift

$$\Delta x = (\mu - 1) t \frac{D}{d}$$

$$\Delta x = (\mu - 1) t \cdot \beta / \lambda$$

Thick of Glass Slab

$$(\mu - 1) t = n\lambda$$

$$t = \frac{n\lambda}{\mu - 1}$$

Wave Optics

Resolving Power of Microscope

$$R_P = \frac{1}{d\theta} = \frac{2\mu \sin\theta}{\lambda}$$

Limit of resolution

$$d\theta = \frac{\lambda}{2\mu \sin\theta}$$

$$d_{min} = \frac{1.22\,\lambda}{\mu \sin\theta}$$

$$\tan\beta = \frac{D}{2y}$$

Resolving Power of Telescope

$$R_P = \frac{1}{d\theta} = \frac{D}{1.22\,\lambda}$$

$$d\theta = \frac{1.22\,\lambda}{D}$$

Diffraction through Single Slit

$$a\sin\theta = n\lambda \quad \rightarrow \text{Minima}$$

$$a\sin\theta = (2n+\tfrac{1}{2})\lambda \quad \rightarrow \text{Max}$$

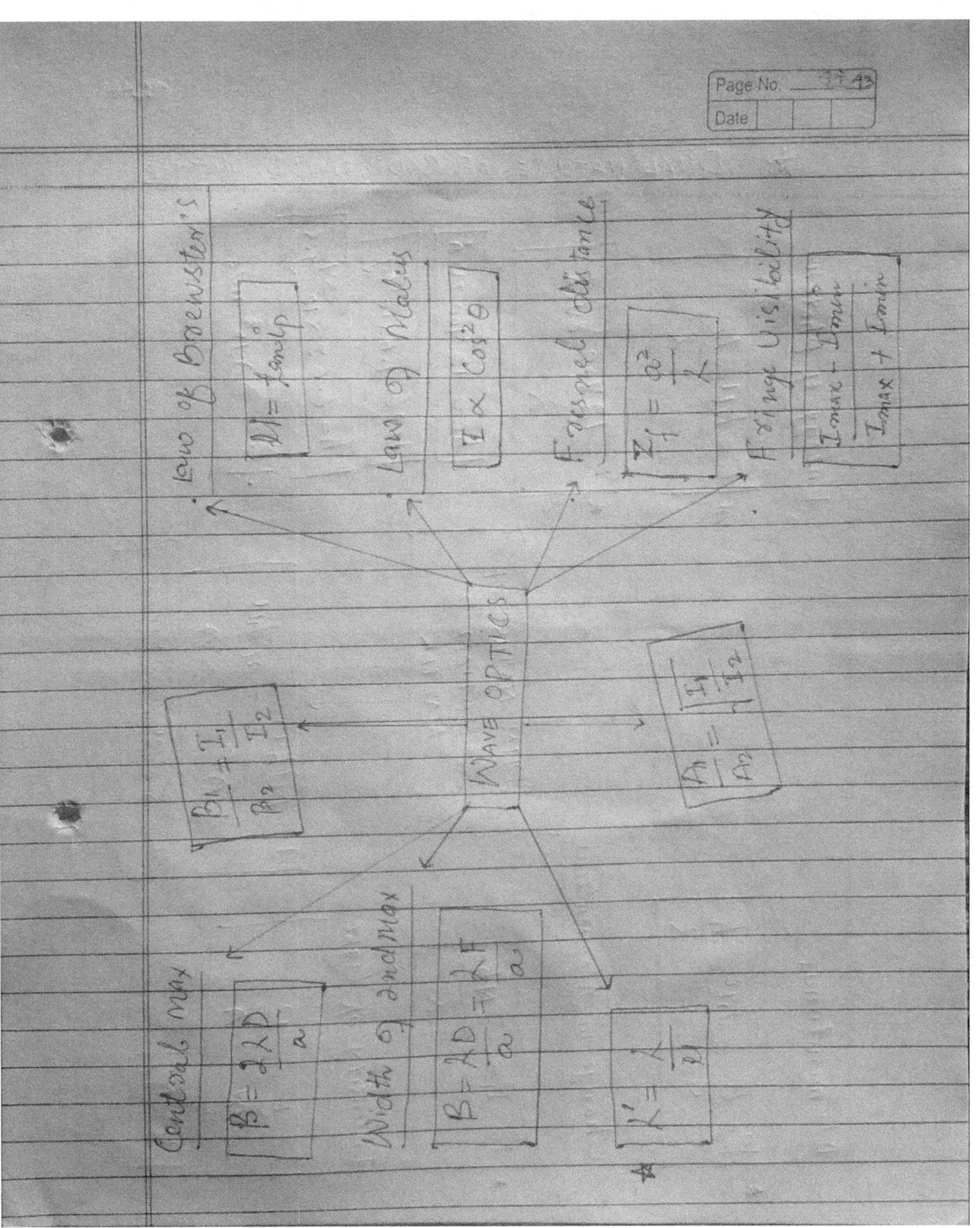

★ DUAL NATURE OF RADIATION & MATTER

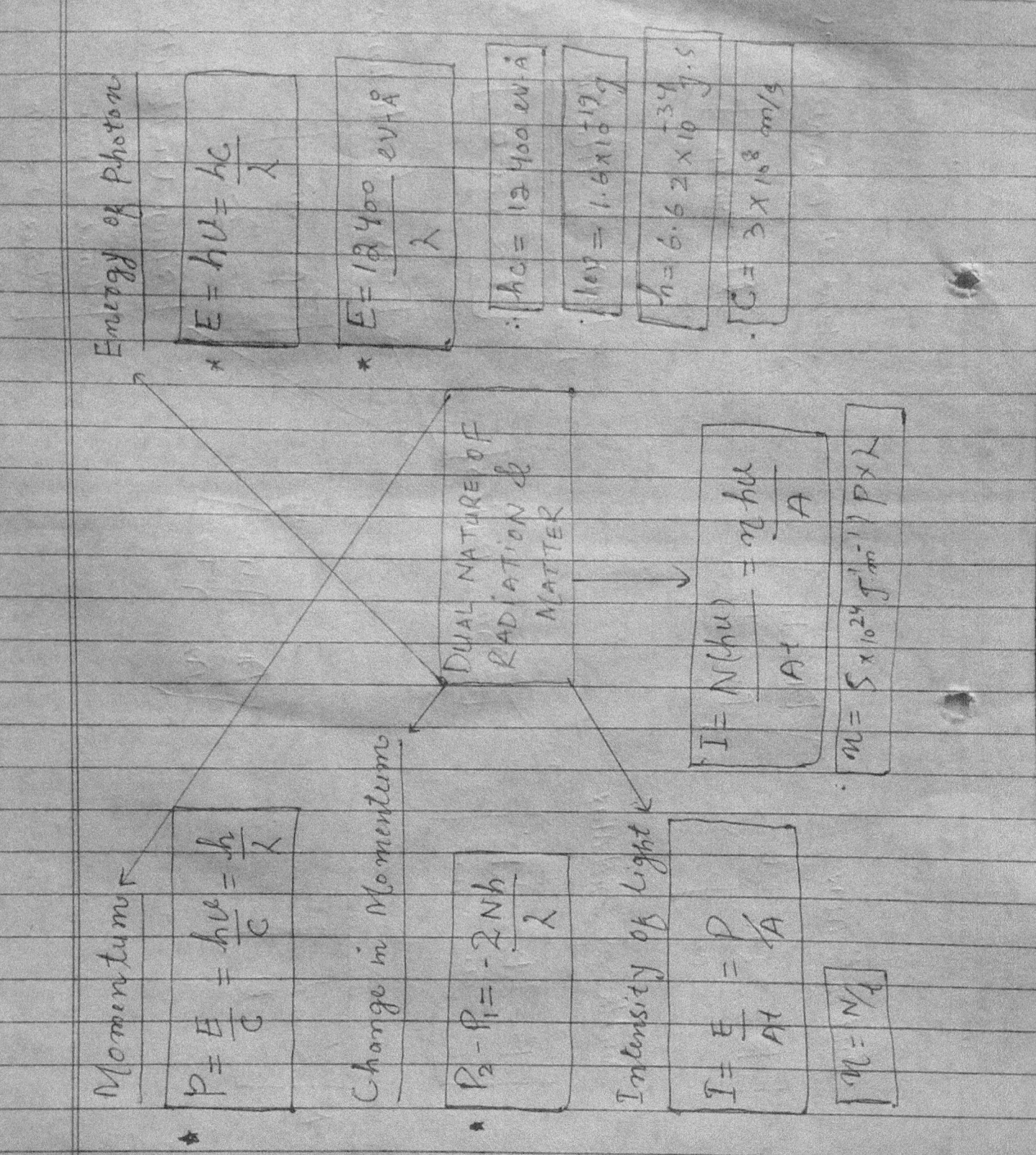

Energy of Photon

* $E = h\nu = \dfrac{hc}{\lambda}$

* $E = \dfrac{12400}{\lambda}\ eV\text{-}Å$

$\therefore hc = 12400\ eV\text{-}Å$
$1\ eV = 1.6 \times 10^{-19}\ J$
$h = 6.62 \times 10^{-34}\ J\text{-}s$
$c = 3 \times 10^8\ m/s$

Momentum

* $P = \dfrac{E}{c} = \dfrac{h\nu}{c} = \dfrac{h}{\lambda}$

Change in Momentum

* $P_2 - P_1 = -\dfrac{2Nh}{\lambda}$

DUAL NATURE OF RADIATION & MATTER

$I = \dfrac{N(h\nu)}{At} = \dfrac{nh\nu}{A}$

$n = 5 \times 10^{24}\ J^{-1}m^{-1}\ P \times \lambda$

Intensity of Light

$I = \dfrac{E}{At} = \dfrac{P}{A}$

$n = N/t$

Photoelectric effects

- $\nu \geq \nu_{threshold}$ fre
  $\lambda \leq \lambda_{Th}$
  ↓
  electrons are ejected

- $\nu \leq \nu_{Th}$ → electrons
  $\lambda > \lambda_{Th}$ cannot be ejected

- Work function (ϕ)
  $$\phi = h\nu_0 = \frac{hc}{\lambda_0}$$

DUAL NATURE OF RADIATION & MATTER

★ Equation of photoelectric effect

- $KE_{max} = h\nu - \phi$
- $KE_{max} = eV_0 = h\nu - h\nu_0$
- $KE_{max} = h(\nu - \nu_0)$

Stopping potential
$$eV_0 = KE_{max} = \frac{1}{2}m v_{max}^2$$

Number of Photons
★ $n = \dfrac{P}{h\nu} = \dfrac{P\lambda}{hc}$

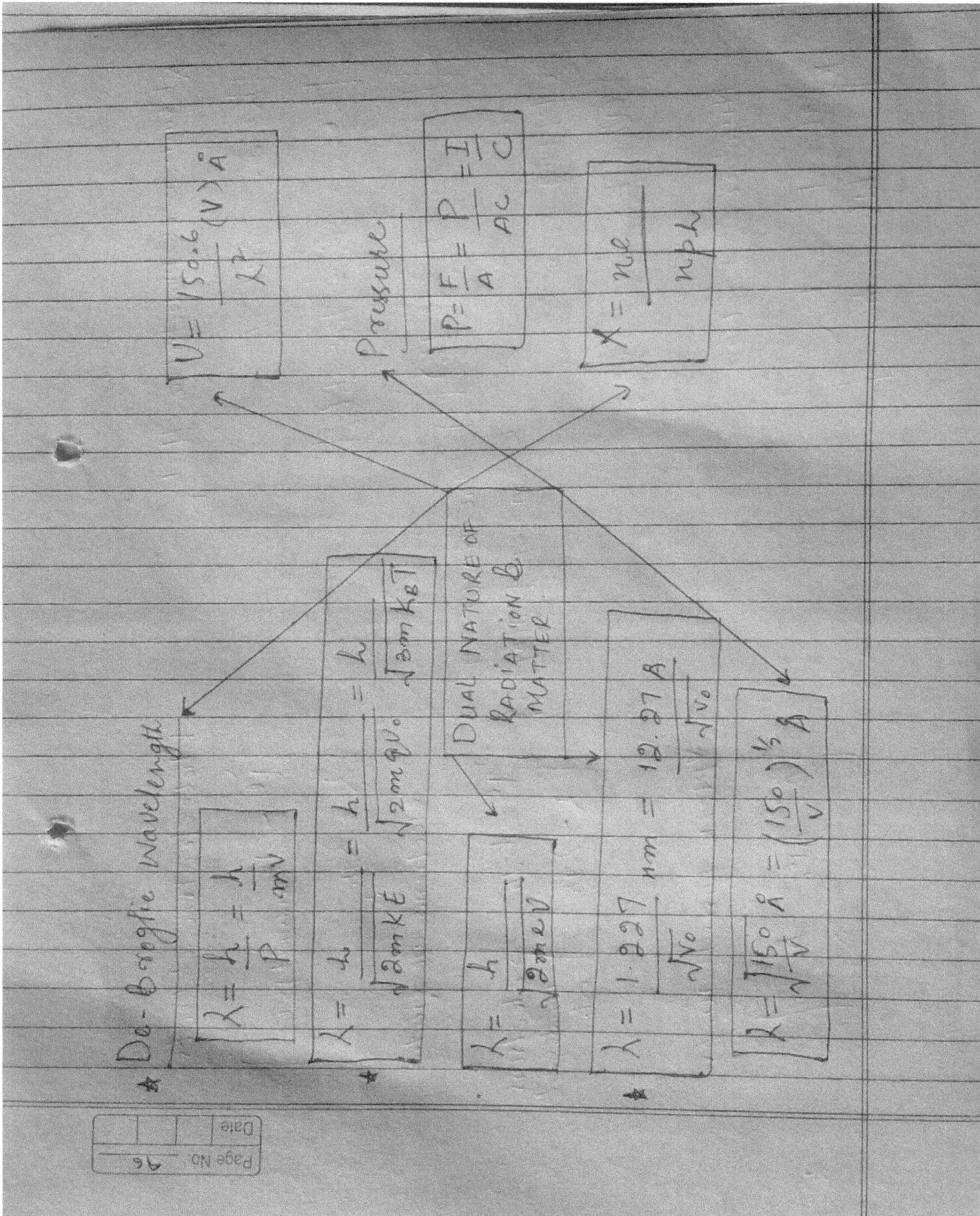

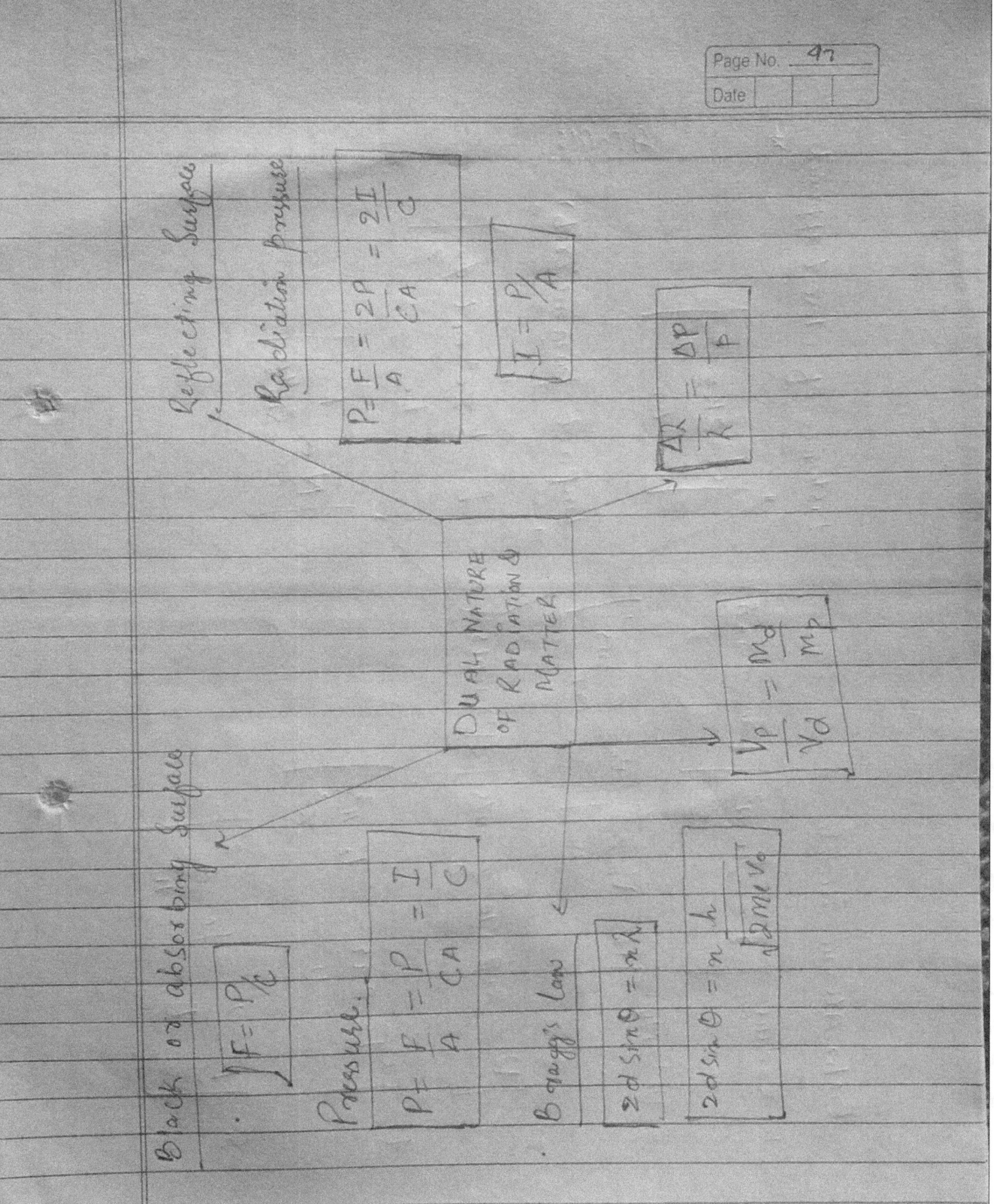

DUAL NATURE OF RADIATION & MATTER

Reflecting Surface

Radiation Pressure

$$P = \frac{F}{A} = \frac{2P}{cA} = \frac{2I}{c}$$

$$I = P/A$$

$$\frac{\Delta x}{k} = \frac{\Delta p}{p}$$

Black or absorbing Surface

$$F = P/c$$

Pressure

$$P = \frac{F}{A} = \frac{P}{cA} = \frac{I}{c}$$

Bragg's Law

$$2d\sin\theta = n\lambda$$

$$2d\sin\theta = n\frac{h}{\sqrt{2mev_0}}$$

$$\frac{v_p}{v_d} = \frac{m_d}{m_p}$$

ATOMS:

Bohr Atomic Model

1. **Radius:** $r = r_0 \dfrac{n^2}{Z}$; $r_0 = 0.53 \text{ Å}$

2. **Velocity:** $v = v_0 \dfrac{Z}{n}$; $v_0 = \dfrac{c}{137} \text{ m/s}$

3. $v = 2.2 \times 10^6 \dfrac{Z}{n} \text{ m/s}$

4. $T = \dfrac{2\pi r}{v} = T_0 \dfrac{n^3}{Z^2}$

5. $f = \dfrac{1}{T} = f_0 \dfrac{Z^2}{n^3}$

Distance of closest approach
$$d = \dfrac{2kZe^2}{E_0}$$

Impact parameter (b)
$$b = \dfrac{kZe^2 \cot \phi/2}{E_0}$$

$$N \propto \dfrac{1}{\sin^4 \theta}$$

ATOMS

Frequency of Emitted Radiation (atom from n_i)

$$\overline{v} = \frac{1}{\lambda} = R\left[\frac{1}{n_1^2} - \frac{1}{n_2^2}\right]$$

$$\frac{1}{\lambda} = R\left[\frac{1}{n_1^2} - \frac{1}{n_2^2}\right] Z^2$$

$R = 10.97 \times 10^6$

→ **Moseley's Law**

$$\sqrt{v} = a(z-b)^2$$

Recoil energy

$$E_r = \frac{P^2}{2m}$$

$$\frac{E_1}{T_2} = \left(\frac{n_2}{n_1}\right)^3$$

- **Kinetic Energy**
  ★ $KE = -E_n$

- **Potential energy**
  ★ $PE = 2E$

Energy (E)

★ $E = \frac{-13.6 Z^2}{n^2}$ eV

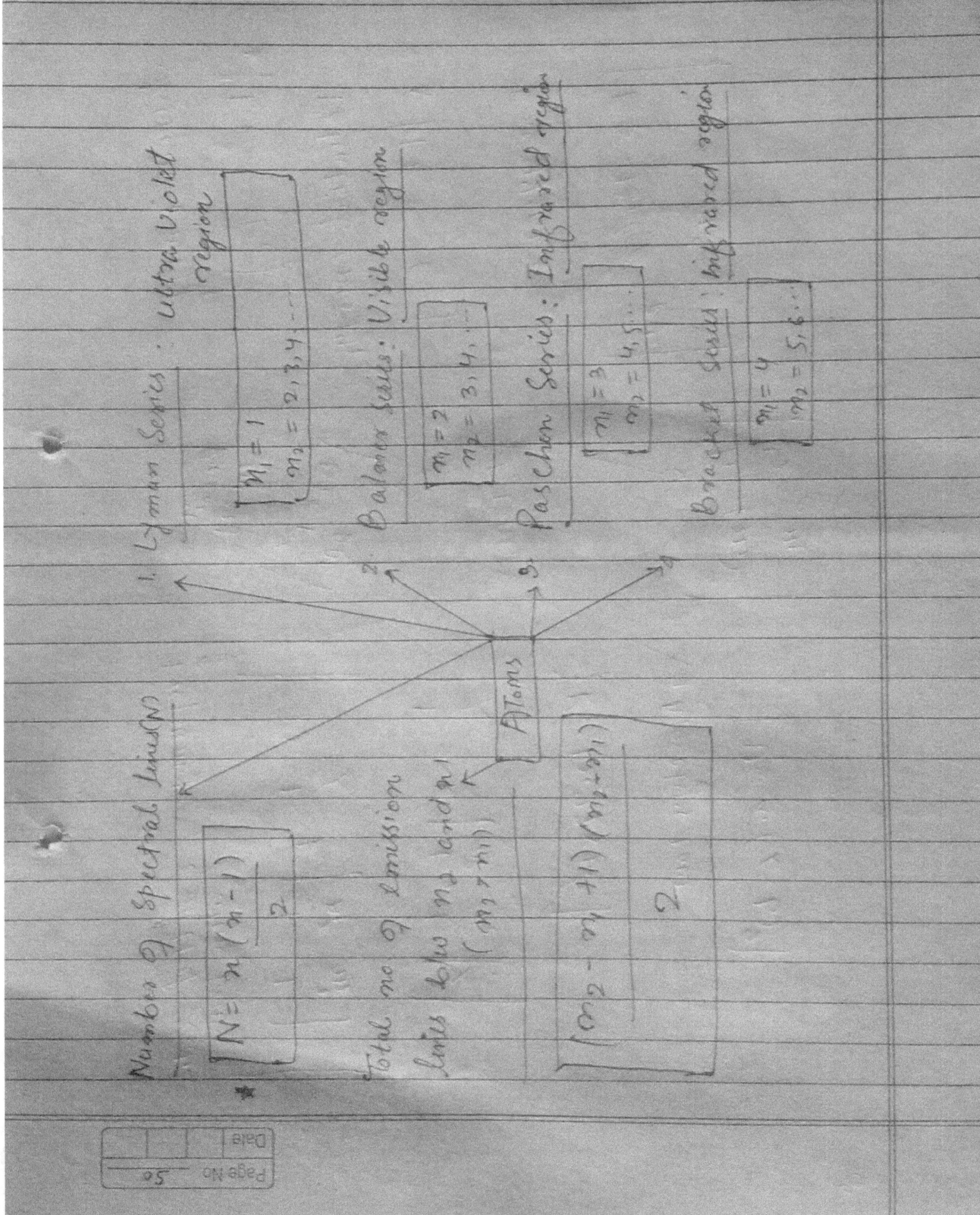

Number of Spectral lines(N)

$$N = \frac{n(n-1)}{2}$$

Total no. of Emission lines b/w no. n_2 and n_1 ($n_2 > n_1$)

$$\frac{(n_2-n_1)(1+n_2-n_1)}{2}$$

ATOMS

1. Lyman Series - Ultra Violet region

$n_1 = 1$
$n_2 = 2, 3, 4, \ldots$

2. Balmer series - Visible region

$n_1 = 2$
$n_2 = 3, 4, \ldots$

3. Paschen Series - Infrared region

$n_1 = 3$
$n_2 = 4, 5, \ldots$

4. Bracket Series - Infrared region

$n_1 = 4$
$n_2 = 5, \ldots \infty$

★ NUCLEI

NUCLEI branches into:

1. Isotopes : Same Z
2. Isobar : Same A
3. Isotone : Same $(A-Z)$

- Atomic Number (Z) → $Z =$ No. of Protons
- Atomic Mass number (A) → $A =$ Proton + Neutrons
- Nucl - mass = $(A-Z)$

Alpha decay:
$$_{Z}X^{A} \rightarrow _{Z-2}Y^{A-4} + _{2}He^{4}$$

Radius of Nucleus
$$R = R_0 A^{1/3}$$
$$R_0 = 1.2 \text{ fm}$$
$$\left(\frac{R_1}{R_2}\right) = \left(\frac{A_1}{A_2}\right)^{1/3}$$

Mass defect:
$$\Delta m = Z m_p + (A-Z) m_n - m_{nucl}$$

Binding energy
$$B.E = \Delta m \cdot c^2$$
$$B.E = \{Z m_p + (A-Z) m_n - m_{nucl}\} c^2$$

Law of Radioactivity

$$\frac{dN}{dt} = \lambda N$$

$$N = N_0 e^{-\lambda t}$$

Half life ($T_{1/2}$)

$$T_{1/2} = \frac{0.693}{\lambda} \approx \frac{0.7}{\lambda}$$

Mean or average life

$$\tau = 1/\lambda = \frac{T_{1/2}}{0.693}$$

$$T = 1.44 T_{1/2}$$

Beta decay

$$Q = (m_x - m_y)c^2$$

$$Q = (m_x - m_y - 2m_e)c^2$$ ← positron

→ Nucleus

[Nuclei]

Activity law

$$R = -\frac{dN}{dt}$$

$$R(t) = R_0 e^{-\lambda t}$$

$$R_0 = \lambda N_0$$

Fraction of nuclei left undecayed after n half lives.

$$\frac{N}{N_0} = \left(\frac{1}{2}\right)^n = \left(\frac{1}{2}\right)^{t/T_{1/2}}$$

or $\quad t = n\, T_{1/2}$

Semiconductor Electronics: Materials, Devices & Simple Circuits

N-type Semiconductor

Pentavalents

Ex. P, As, Sb, Bi

$$n_e \gg n_h$$

electron → Majority
hole → Minority

P-type Semiconductor

Trivalents

Ex. B, Al, In, Ga

$$n_h \gg n_e$$

hole → Majority
electron → Minority

$$n_i^2 = n_e \cdot n_h$$

Semiconductor

Dynamic Resistance

$$r_d = \frac{\Delta V}{\Delta I}$$

Forward Bias

$$V_P > V_N$$

N → −ve
P → +ve

Reverse Bias

$$V_N > V_P$$

N → +ve
P → −ve

Strength of Junction field

$$E = \frac{\Delta V}{d}$$ $$E = \frac{V}{dL}$$

Collector: Largest size
Moderately doped

Emitter: Highly doped | Semiconductor
Moderately sized

Base: Lightly doped,
Small sized

Half Wave Rectifier

$$I_{rms} = \frac{I_o}{2}$$

$$I_{dc} = \frac{I_o}{\pi}$$

$$\eta_{max} = 40.6\%$$

Full Wave Rectifier

$$I_{rms} = \frac{I_o}{\sqrt{2}}$$

$$I_{dc} = \frac{2 I_o}{\pi}$$

$$\eta_{max} = 81.2\%$$

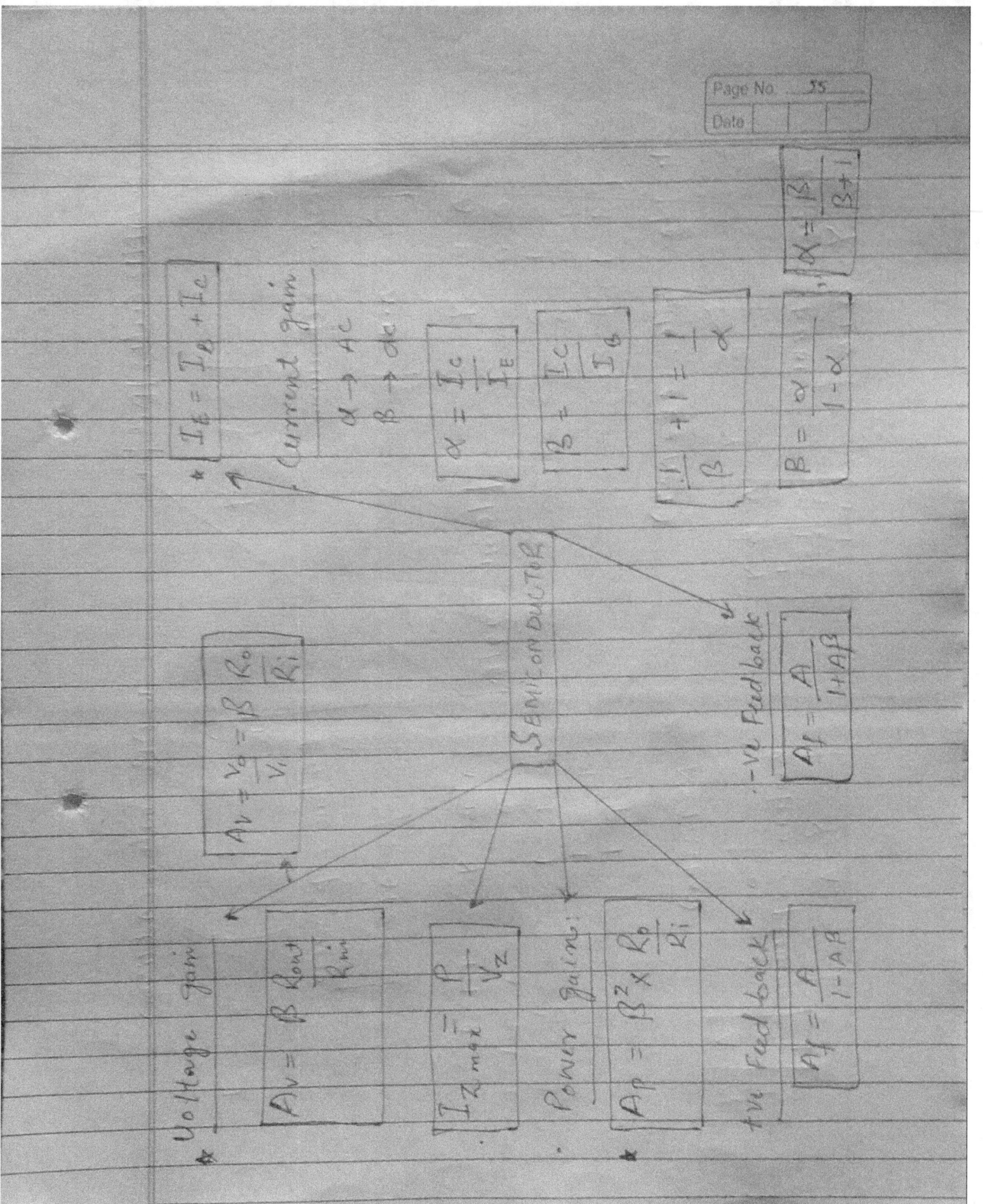

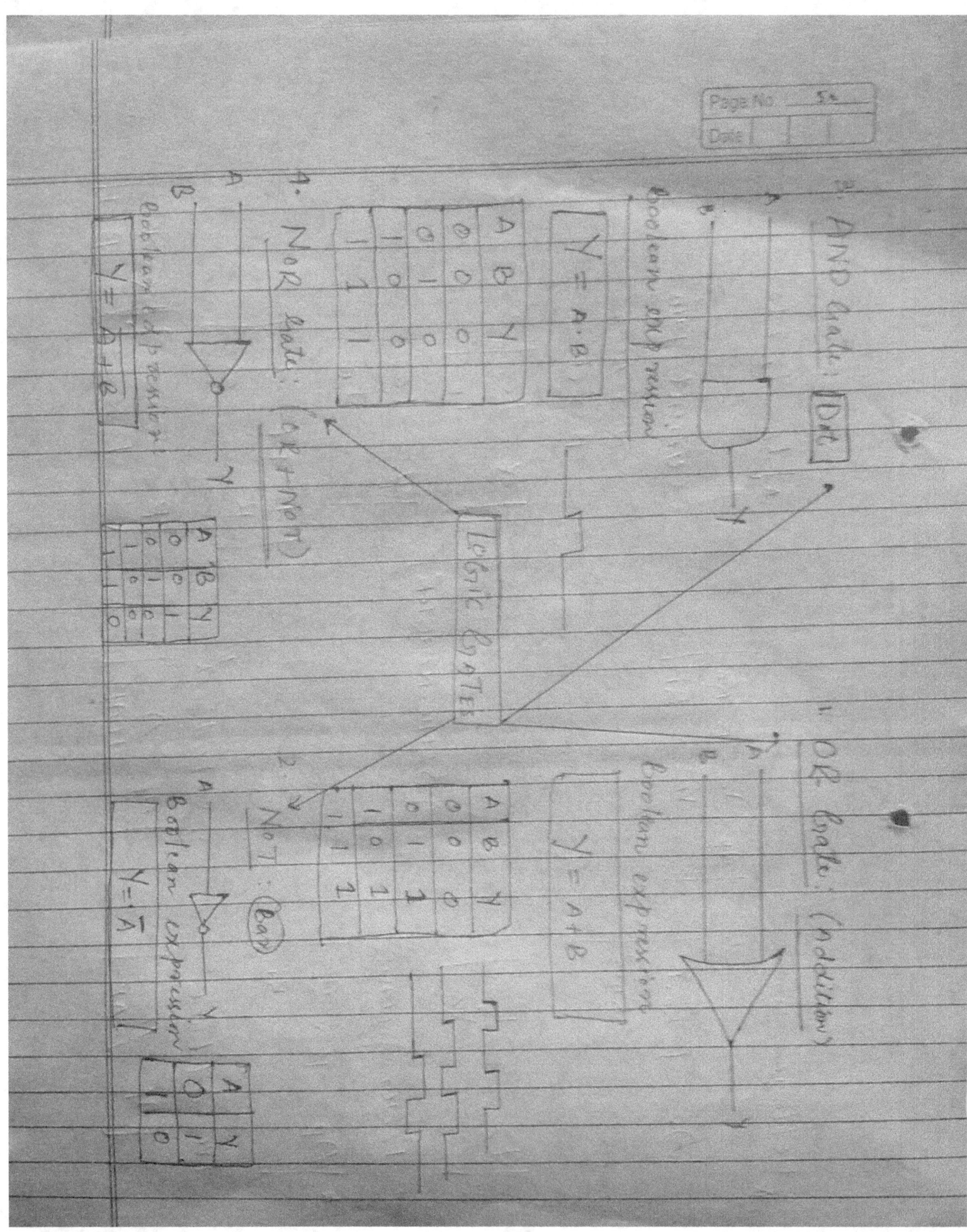

LOGIC GATES

3. AND Gate: (Dot)

Boolean expression
$$Y = A \cdot B$$

A	B	Y
0	0	0
0	1	0
1	0	0
1	1	1

1. OR Gate: (addition)

Boolean expression
$$Y = A + B$$

A	B	Y
0	0	0
0	1	1
1	0	1
1	1	1

4. NOR Gate: (OR + NOT)

Boolean expression:
$$Y = \overline{A + B}$$

A	B	Y
0	0	1
0	1	0
1	0	0
1	1	0

2. NOT Gate

Boolean expression
$$Y = \overline{A}$$

A	Y
0	1
1	0

5. NAND gate: (AND + NOT)

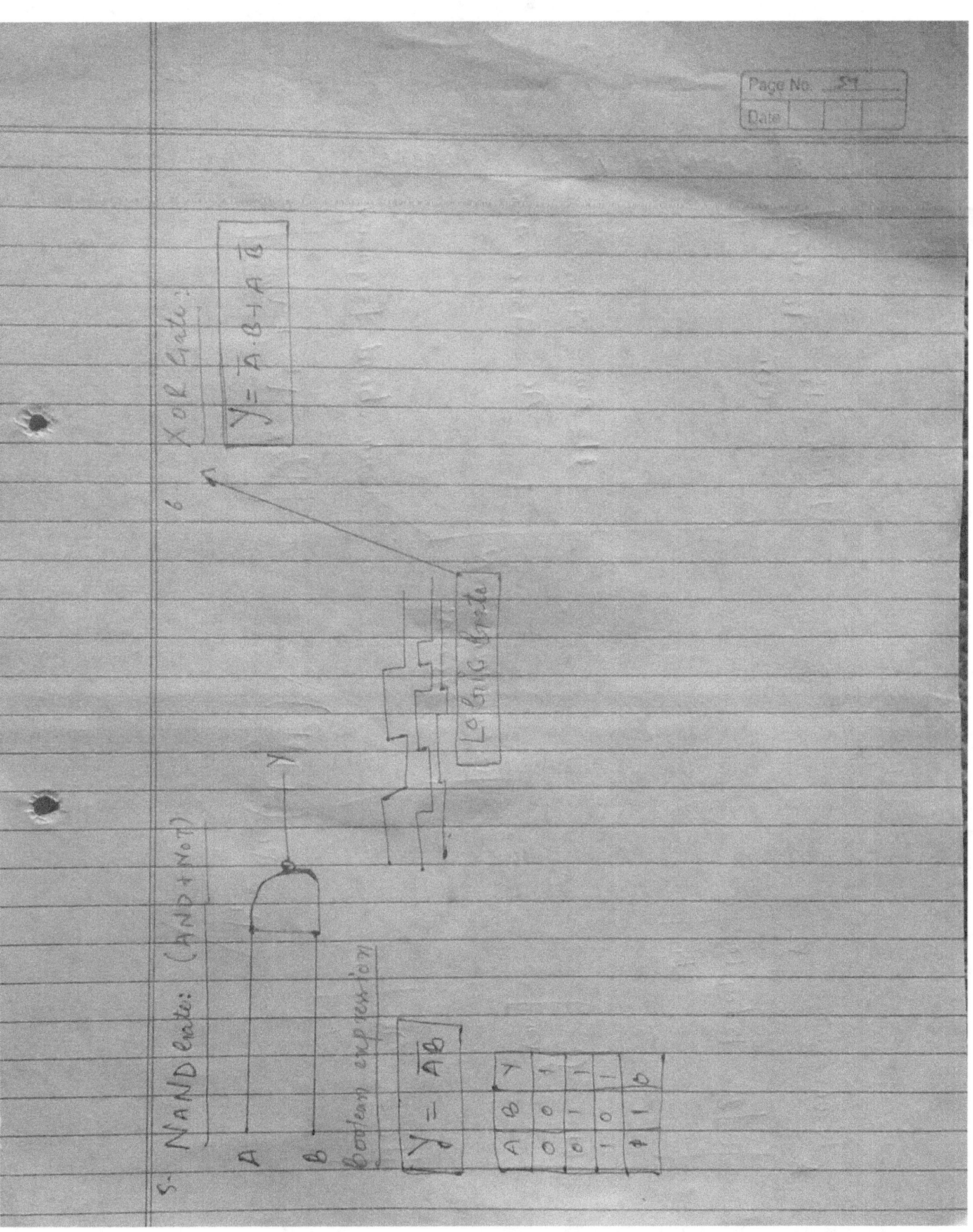

Boolean expression

$$y = \overline{AB}$$

A	B	Y
0	0	1
0	1	1
1	0	1
1	1	0

Logic Gate

6. XOR Gate:

$$y = \overline{A}\cdot B + A\cdot \overline{B}$$

Communication System

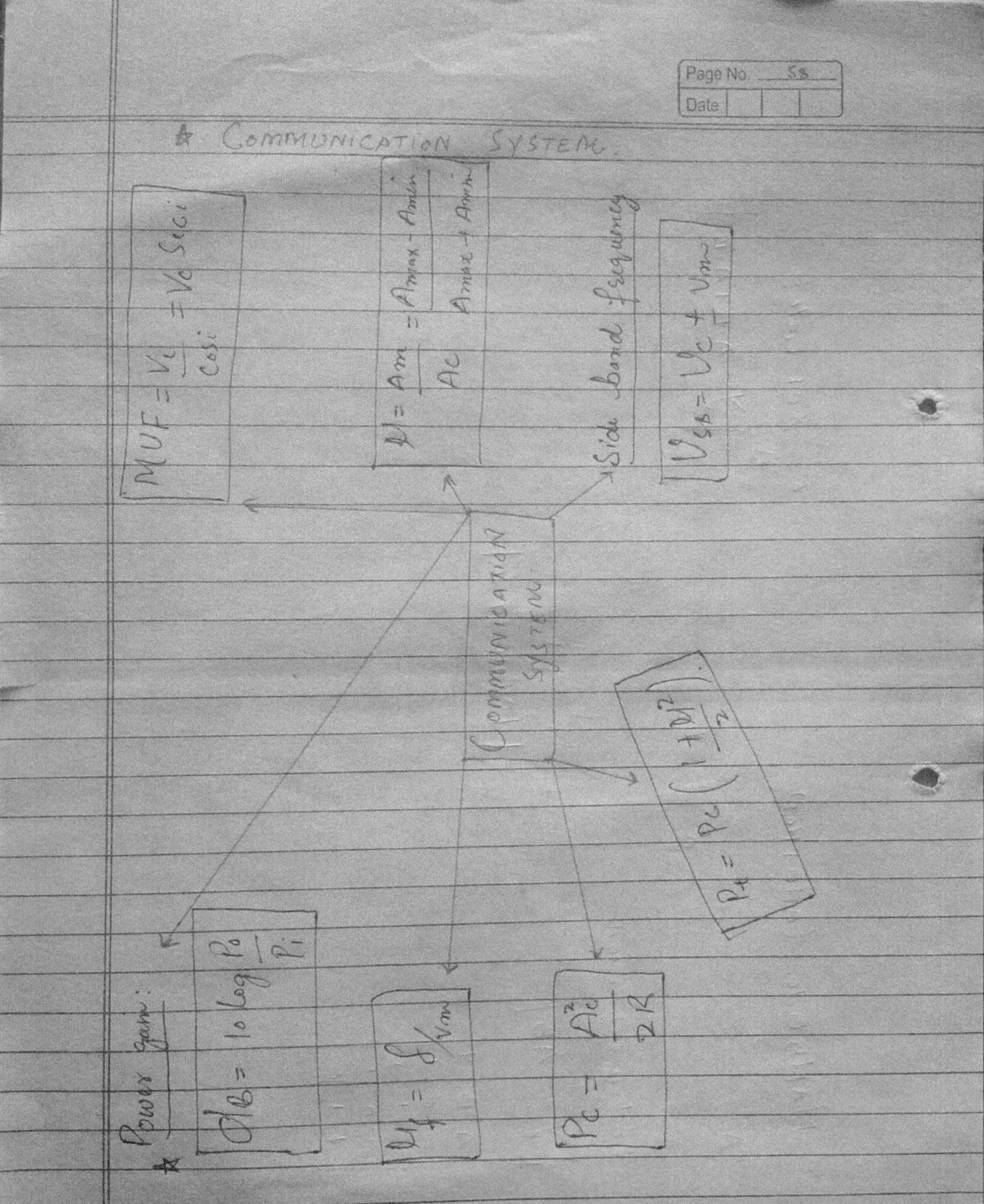

$$MUF = \frac{V_c}{\cos i} = V_0 \sec i$$

$$\mu = \frac{A_m}{A_c} = \frac{A_{max} - A_{min}}{A_{max} + A_{min}}$$

Side band frequency

$$V_{SB} = V_c \pm V_m$$

Communication System

$$P_t = P_c\left(1 + \frac{\mu^2}{2}\right)$$

★ Power gain:

$$dB = 10 \log \frac{P_0}{P_i}$$

$$\mu\% = \frac{\mu}{1} \times 100$$

$$P_c = \frac{A_c^2}{2R}$$